Developing a Methodology for Risk-Informed Trade-Space Analysis in Acquisition

Craig A. Bond, Lauren A. Mayer, Michael E. McMahon, James G. Kallimani, Ricardo Sanchez

Prepared for the United States Army

Approved for public release; distribution unlimited

For more information on this publication, visit www.rand.org/t/rr701

Library of Congress Cataloging-in-Publication Data is available for this publication.

ISBN 978-0-8330-8764-5

Published by the RAND Corporation, Santa Monica, Calif.

© Copyright 2015 RAND Corporation

RAND® is a registered trademark.

Support RAND

Make a tax-deductible charitable contribution at
www.rand.org/giving/contribute

www.rand.org

Preface

This report documents the results of the project "Developing a Methodology Framework for Conducting Risk-Informed Trade Space Analyses." The primary objective of this study was to develop a framework and acquisition risk-assessment tool that conducts schedule, funding, and performance trades for a given materiel alternative and that provides the consequences of these trades in terms of quantifiable risk. The U.S. Army has a process by which it can assess the risk and performance of a given analysis-of-alternatives (AoA) alternative; however, it does not have a robust quantitative framework to link the outcome of the AoA risk (cost, schedule, and technical) and performance assessments to consistently assess the impact of a schedule, funding, or performance trade among alternatives.

This report describes the methodology for the risk-informed trade space jointly developed by the U.S. Army Materiel Systems Analysis Activity (AMSAA) Risk Team and RAND Corporation researchers and the first iteration of the associated Risk-Informed Trade Analysis Model (RTRAM). The framework and model allow users to investigate multidimensional trade-offs both within and between weapon systems prior to production using elements of system engineering, production economics, and risk analysis to functionally and probabilistically relate performance, schedule, and cost outcomes and their related uncertainties holistically and understandably. The framework should be useful to those who wish to compare joint performance, schedule, and cost outcomes and uncertainties in an acquisition environment such as that described by an AoA. It is not intended to provide recommendations for a particular system over others. Rather, it is a decision-support methodology that allows users to investigate multidimensional trade-offs both within and between weapon systems prior to production.

This research was sponsored by Randolph Wheeler, AMSAA-WR, and conducted within the RAND Arroyo Center's Force Development and Technology Program. RAND Arroyo Center, part of the RAND Corporation, is a federally funded research and development center sponsored by the United States Army.

The Project Unique Identification Code (PUIC) for the project that produced this document is RAN136454.

Contents

Figures

Tables

Summary

One of the top priorities established by the Secretary of the Army is to ensure a highly capable Army within fiscal constraints. To acquire and modernize the U.S. Army, the annual Army Equipment Modernization Strategy requires Army leadership to make decisions regarding acquisition programs that will best serve the warfighter in a fiscally constrained environment. A U.S. Army Materiel Systems Analysis Activity (AMSAA)–led Army Risk Team was formed in March 2011, at the direction of Army leadership, to develop standard methodologies for conducting independent risk assessments (performance, schedule, and cost) to support analyses of alternatives (AoAs) and other major Army acquisition studies. The Weapon Systems Acquisition Reform Act of 2009 (Pub. L. No. 111-23) has also driven increased analysis to support AoAs, including risk assessments and trades among cost, schedule, and performance. It is essential to have effective methods to conduct risk-informed trade-space analyses to better inform senior Army leaders and support difficult decisions about materiel alternatives.

The primary objective of this study was to develop a framework and acquisition risk-assessment tool that conducts schedule, funding, and performance trades for a given materiel alternative and that provides the consequences of these trades in terms of quantifiable risk.

Standard practice across the Army (and, to a lesser extent, the U.S. Department of Defense [DoD] as a whole) is to treat technological performance, schedule, and cost estimation and risk as virtually independent dimensions. That is not to say that, for example, cost analysts do not incorporate technical or schedule considerations; rather, assumed input distributions on each work breakdown structure cost element represent these dimensions implicitly. Although informative, the resultant cost distribution does a poor job of helping decisionmakers make choices about trade-offs and risk management because there is no traceable, structural link between technological performance, schedule, and cost outcomes.

It is in this spirit that the AMSAA Risk Team and RAND researchers jointly developed a new decision-support methodology for risk-informed trade-space analysis in weapon-system acquisition and the first iteration of the Risk-Informed Trade Analysis Model (RTRAM). The framework and model allow users to investigate multidimensional trade-offs both within and between weapon systems prior to production using elements of system engineering, production economics, and risk analysis to functionally and probabilistically relate performance, schedule, and cost outcomes and their related uncertainties holistically and understandably. In the model, the technology-development process is conceptualized as a physical system consisting of a portfolio of technologies with associated technical capabilities, and the completion of each technology is stochastic (i.e., a discrete random variable). As a result, the final system's performance characteristics are stochastic. In addition, the time of technology development is also stochastic and, in part, drives the overall cost of the system.

A novel feature of the model is the incorporation of technology-specific courses of action, or risk-mitigation behaviors, which take place in the event that the technology is not developed by the milestone date (e.g., allowing for performance degradation, schedule slippage, or increased investment). This allows for a quantitative evaluation of potential risk-mitigating actions across the multidimensional output space. The framework should be useful to those who wish to compare a set of alternatives in an acquisition environment, such as that completed in an AoA, using joint performance, schedule, and cost outcomes and uncertainties.

How the Project Was Performed

RAND researchers worked closely with the AMSAA Risk Team to first understand the current processes and work to determine current AMSAA methodologies, processes, and results; identify key analysis capability gaps; and develop the risk-adjusted trade-space methodology and RTRAM tool. Through the course of the research, the partner teams met regularly (approximately monthly) to share, vet, and clarify ideas; develop algorithms; and discuss details of the methodology and RTRAM tool. This resulted in first a model framework and later the development of a baseline risk tool for AMSAA to take forward.

The reader should thus consider the framework and tool documented herein to be outputs jointly developed by the AMSAA Risk Team and the RAND team.

Acknowledgments

We wish to thank a great many individuals at multiple organizations who worked with us to perform this research. We would like to thank Randolph Wheeler, Suzanne Singleton, Thomas Bounds, Andrew Clark, John Nierwinski, and the rest of the team at AMSAA for providing timely and comprehensive guidance and feedback. We also appreciate the support we received from Alison Tichenor at the Office of the Deputy Assistant Secretary of the Army for Cost and Economics; Frank Decker, Cindy Noble, and Michele Wolfe at the U.S. Army Training and Doctrine Command Analysis Center at Fort Leavenworth; and Everett Johnson at the U.S. Army Training and Doctrine Command Analysis Center at White Sands Missile Range. Elaine Simmons from the Office of Cost Assessment and Program Evaluation provided feedback for the project. During the review process, Frank Camm of RAND and Paul Garvey of the MITRE Corporation provided thoughtful comments and suggestions that greatly improved the report. Paul Dreyer of RAND provided a thorough quality assurance review of RTRAM. We are also grateful to Christopher G. Pernin and Timothy M. Bonds, both of RAND, for enabling the research and providing feedback on the work. We would also like to express our gratitude to Lauren Varga, Alexander Chinh, Maggie Snyder, Elizabeth Cole, Andria Tyner, and Michelle Horner for providing administrative support.

Abbreviations

AMSAA	U.S. Army Materiel Systems Analysis Activity
AoA	analysis of alternatives
APUC	average procurement unit cost
ARCIC	Army Capabilities Integration Center
ASA(FM&C)	Assistant Secretary of the Army for Financial Management and Comptroller
CAPE	Office of Cost Assessment and Program Evaluation
CDD	concept design document
cdf	cumulative distribution function
CDR	critical design review
CER	cost-estimating relationship
COA	course of action
DASA-CE	Deputy Assistant Secretary of the Army for Cost and Economics
DoD	U.S. Department of Defense

EMD	engineering and manufacturing development
FOC	full operational capability
FRP	full-rate production
HQDA	Headquarters, Department of the Army
IRL	integration readiness level
KPP	key performance parameter
KT	key technology
MC	Monte Carlo
MDAP	major defense acquisition program
MRL	manufacturing readiness level
MS	milestone
ODASA-CE	Office of the Deputy Assistant Secretary of the Army for Cost and Economics
OSD	Office of the Secretary of Defense
OUSD(AT&L)	Office of the Under Secretary of Defense for Acquisition, Technology, and Logistics
PAUC	program acquisition unit cost
pdf	probability distribution function
PDR	preliminary design review
PM	program manager
RDECOM	U.S. Army Research, Development and Engineering Command
RDT&E	research, development, test, and evaluation

RTRAM	Risk-Informed Trade Analysis Model
SDT	schedule delivery time
SME	subject-matter expert
SRDDM	Schedule Risk Data Decision Methodology
TOA	time of arrival
TRA	Technology Readiness Assessment
TRAC	U.S. Army Training and Doctrinc Command Analysis Center
TRAC-WSMR	U.S. Army Training and Doctrine Command Analysis Center at White Sands Missile Range
TRADOC	U.S. Army Training and Doctrine Command
TRL	technology readiness level
UI	user interface
VBA	Visual Basic for Applications
WBS	work breakdown structure

CHAPTER ONE

Introduction

Background

The Weapon Systems Acquisition Reform Act of 2009 mandates, at a minimum, that analysis-of-alternatives (AoA) study guidance include

> (1) full consideration of possible trade-offs among cost, schedule, and performance objectives for each alternative considered; and (2) an assessment of whether or not the joint military requirement can be met in a manner that is consistent with the cost and schedule objectives recommended by the Joint Requirements Oversight Council. (Pub. L. 111-23, Title II, § 201[b][5][d][2])

Under the act, the director of the Office of Cost Assessment and Program Evaluation (CAPE) is the primary official within the U.S. Department of Defense (DoD) for formulating AoA study guidance for major defense acquisition programs (MDAPs) and assessing the performance of those analyses (Pub. L. 111-23, Title I, § 101). In this context, CAPE has communicated to its stakeholders that it values objective and independent inquiry into the costs and capabilities of options, with clear, straightforward analysis that highlights and explains key trade-offs between performance, schedule, and cost outcomes (CAPE, 2013). It also encourages sensitivity analysis, including analysis of worst-case scenarios, and economy in reporting AoA results (CAPE, 2013). Less valued is analysis that makes the case for a preferred solution, treats key performance parameters (KPPs) as hard constraints on alternatives, and focuses on performance and capabilities without highlighting associated risks (CAPE, 2013).

The uncertainty and risk associated with the acquisition process is well documented.[1] For example, a 2013 report on acquisition-system performance from the Office of the Secretary of Defense (OSD) explicitly states that "[a]cquisition is about risk management—not certainties," especially for major weapon systems that involve significant research and development (Office of the Under Secretary of Defense for Acquisition, Technology, and Logistics [OUSD(AT&L)], 2013, p. 109). Arena et al. (2006) and references therein provide an excellent summary of RAND research on the subject of weapon cost growth since the 1950s, with total average cost overages relative to estimates at MS B for weapon systems estimated at approximately 46 percent. Over the past ten years, another estimate puts median program RDT&E cost growth at 5 to 18 percent, suggesting that this problem has been persistent across time (OUSD[AT&L], 2013; Younossi, Arena, et al., 2007).

These overages are due to the uncertainty inherent in the acquisition and technology-development processes. The sources of this uncertainty have been classified by Bolten et al. (2008) into four major categories: (1) errors in estimation and planning; (2) decisions by the government, including changes in requirements and other programmatic changes; (3) financial matters, including changes in the macroeconomic environment; and (4) miscellaneous sources. They found that decisions by the government accounted for approximately two-thirds of total cost growth for 35 select mature programs, while estimating errors accounted for approximately 25 percent of overages.

---

[1] The terms *uncertainty* and *risk* can have different meanings across disciplines and contexts. Here, *uncertainty* refers to the fact that outcomes are not known before observation (e.g., total research, development, test, and evaluation [RDT&E] costs of a program when analyzed prior to Milestone [MS] C are uncertain). In the acquisition process, milestones are specific requirements that must be met before the program can continue. Reaching MS C indicates a readiness for production and deployment. *Risk* refers to the joint probability and consequence of adverse outcomes and is measured relative to some objective (e.g., the risk of RDT&E costs exceeding a particular target involves the probability of exceeding and the consequences of doing so). We do not consider cases in which a probability distribution is completely unknown (sometimes termed *true uncertainty* or *ambiguity*).

Within the literature on cost growth, there appears to be consensus that technical, schedule, and cost risks are interconnected, with technical and schedule outcomes feeding into resultant costs. For example, OSD has stated, "Performance (good or bad) in planned defense acquisition is intertwined with cost and schedule implications . . . ." (OUSD[AT&L], 2013, p. 109). Bolten et al. (2008) decomposed their errors in the estimation and planning category into cost, schedule, and technical components, implying that all three feed into ultimate cost outcomes. Younossi, Lorell, et al. (2008) states,

> Technical risks, such as immature technologies or a compressed testing schedule, lead to technical difficulties that could eventually result in failures in meeting the technical performance. As a result, redesigns and rework may be required, which slow the progress of the program and cause schedule slips and cost growth. (p. 45)

Despite this recognition, and perhaps because of the complexity involved, standard practice across the Army (and, to a lesser extent, DoD as a whole) is to treat technical, schedule, and cost estimation and risk as virtually independent dimensions. For example, current state of practice in cost risk analysis is essentially the use of cost-estimating relationships (CERs), often obtained through statistical analysis at the project or subcomponent level, augmented by Monte Carlo (MC) simulations based on input distributions derived from subject-matter experts (SMEs) or perhaps other data sources (Arena et al., 2006). However, the details vary considerably across organizations and implementation. Of particular interest, Arena et al. (2006) wrote, "This approach does not easily allow for including the effect of schedule variation on costs" (p. 66). They explained that the MC simulation procedure is performed at the work breakdown structure (WBS) level, and neither the WBS nor most CERs include schedule as explicit parameters.[2] That is not to say that cost analysts do not incorporate techni-

---

[2] Interestingly, a series of RAND research reports in the 1970s describe the time-of-arrival (TOA) methodology for improving CERs in aircraft turbine engines. Essentially, a statistical relationship is estimated for TOA as a function of technical characteristics, which is then

cal or schedule considerations; rather, these dimensions are *implicitly* represented by the assumed input distributions on each cost element. Even using statistical methods that explicitly recognize the jointness of schedule and cost outcomes, such as the procedures documented in Garvey (2000) and Covert (2013), the decisionmaker is not provided with information "about *which* risks are covered, how they are covered, to what extent they are covered, and how to manage them" (Arena et al., 2006, p. 67, emphasis in original). In other words, when used as stand-alone outputs, marginal or joint probability distributions do a less-than-adequate job of helping decisionmakers make choices about trade-offs and risk management. Indeed, Arena et al. (2006) suggests that "it would be desirable to keep the advantages of a probabilistic approach while making more transparent both the results and how they depend on specific hazards" (p. 67).

It is in this spirit that the U.S. Army Materiel Systems Analysis Activity (AMSAA) Risk Team and RAND researchers jointly developed the following methodology for risk-informed trade-space analysis in weapon-system acquisition and the first iteration of RTRAM. The framework combines elements of system engineering, production economics, and risk analysis to functionally and probabilistically relate performance, schedule, and cost outcomes and their related uncertainties holistically and understandably. The technology-development process is conceptualized as one in which the physical system is a portfolio of technologies with associated performance capabilities, and the completion of each technology at a user-defined MS date is a (discrete) random variable. As such, the final system's performance characteristics are random. In addition, the time of technology development is also random and, in part, drives the overall cost of the system. Model inputs are conceptualized as coming from SMEs and include technology-specific distributions describing technology, integration, and manufac-

---

related to cost and schedule changes (Alexander and Nelson, 1972; Shishko, 1973; Nelson and Timson, 1974; Nelson, 1975). Nelson (1975) states, "This approach is unique in the sense that it allows time, in terms of schedule, to be introduced explicitly in the tradeoff and risk analysis" (p. 1). Unlike the Risk-Informed Trade Analysis Model (RTRAM), which is a stochastic structural model, this methodology is statistical in nature.

turing readiness levels; perhaps performance consequences of development failures; and fixed and variable-cost parameters.[3]

In a departure from previous analyses, we incorporate technology-specific courses of action (COAs), or behaviors, that are assumed to take place in the event that the technology is not developed at the milestone date. For example, one might assume that a particular developmental technology can be replaced by a lesser-performing, already-developed substitute or, if that technology is of critical importance, that schedule slippage might be allowed to occur. By analyzing alternative COAs and their effects on the resultant probability distributions estimated for performance, schedule, and cost, decisionmakers have a means to understand the implications of certain risk-mitigating actions. Technology, schedule, or cost trades can be examined between or within individual systems.

The framework should be useful to those who wish to compare joint performance, schedule, and cost outcomes and uncertainties in an acquisition environment such as that described by an AoA. We do not intend for it to provide recommendations for a particular system over others. Rather, it is a decision-support methodology that allows users to investigate multidimensional trade-offs both within and between weapon systems before production. Unlike more-traditional analyses that focus on one dimension at a time, the risk-informed trade-space methodology attempts to meet decisionmakers' needs by providing "a realistic assessment of the program risks (technical, schedule, or cost) presented to them at key decision points in the life of each program" (Arena et al., 2006). Furthermore, RTRAM's user interface provides multilayered decision support. Initially, users can use RTRAM to explore the entire decision trade space, choosing the most preferred alternatives. These preferred alternatives may then be displayed in RTRAM as a higher-level comparison in a format digestible to upper-level decisionmakers. We do not intend the proposed methodology to be a replacement for a detailed analysis of technological, schedule, or cost risk. Rather, it is complementary to these products, with the objec-

---

[3] Although this is the underlying assumption throughout this document, the methodology itself is considerably more general.

tive of providing ballpark estimates of the trade-offs in potential outcomes in each dimension, either between or within systems.

Objective of This Study

The primary objective of this study was to develop a framework and acquisition risk-assessment tool that conducts schedule, funding, and performance trades for a given materiel alternative and that provides the consequences of these trades in terms of quantifiable risk.

Research Approach

We worked closely with the AMSAA Risk Team to determine current AMSAA methodologies, processes, and results; identify key analysis capability gaps; and develop the risk-adjusted trade-space methodology and RTRAM tool. Through the course of the research, the partner teams met regularly (approximately monthly) to share, vet, and clarify ideas; develop algorithms; and discuss details of the methodology and RTRAM tool. The reader should thus consider the framework and tool documented herein to be outputs jointly developed by the AMSAA Risk Team and RAND researchers.

Outside of these working meetings, we reviewed additional risk methodologies used across DoD by reviewing relevant documents produced by the appropriate organizations, including CAPE, Office of the Deputy Assistant Secretary of the Army for Cost and Economics (ODASA-CE), U.S. Army Training and Doctrine Command Analysis Center (TRAC), OSD, and the Congressional Budget Office and meeting with representatives of a subset of these groups, as documented in Table 1.1. We reviewed documents and briefs related to users of AoAs and other risk analyses (e.g., CAPE) to understand user needs and utilized peer-reviewed research by RAND researchers and in the academic literature to investigate and develop the risk-informed trade-space framework.

Table 1.1
In-Person and Teleconference Meetings for the Project

Date	Organization or Event
October 11, 2012	AMSAA
October 15 to October 19, 2012	Armed Aerial Scout Risk Workshop
December 3, 2012	AMSAA
January 24, 2013	AMSAA
March 1, 2013	CAPE
March 13, 2013	ODASA-CE
April 2, 2013	AMSAA
April 4, 2013	AMSAA
April 5, 2013	TRAC
April 10, 2013	AMSAA
April 16, 2013	AMSAA
April 26, 2013	AMSAA
May 7, 2013	AMSAA
May 15, 2013	AMSAA, TRAC-WSMR
May 22, 2013	AMSAA
June 4, 2013	AMSAA, NAVAIR
June 21, 2013	AMSAA
July 17, 2013	AMSAA, ODASA-CE
August 7, 2013	AMSAA
September 4, 2013	AMSAA
September 23, 2013	AMSAA

NOTE: TRAC-WSMR = TRAC White Sands Missile Range.
NAVAIR = Naval Air Systems Command.

Structure of the Model

The complexity of the problem required a multidimensional and intricate model that we detail in this report. Although it is complex, we can explain the logic underlying the model with these simple steps:

1. *Identify technologies.* For each alternative weapon system, identify a set of key technologies that are critical to the system's success.
2. *Determine a schedule.* For each key technology (KT), determine a triangular schedule distribution that represents the probability that this technology will be delivered (i.e., meet MS C criteria) by a specific time (the milestone date).
3. *Determine consequences.* For each KT, determine the consequence of that technology not being delivered by the milestone date.
4. *Choose COAs.* Choose a COA that can be used to mitigate the risk of consequence for each technology.
5. *Define the schedule–cost relationship.* Define a relationship between schedule and cost (such that cost is a function of schedule).
6. *Draw a schedule date.* For each technology, use MC methods to draw a schedule date from the triangular distribution. If schedule date is after MS C, apply COAs as necessary.
7. *Calculate a schedule estimate.* Use technology-specific delivery dates to aggregate to a weapon-system delivery date. This provides a schedule estimate.
8. *Calculate a performance estimate.* For each technology, assign the consequence associated with the technology-specific delivery dates. Aggregate to a weapon-system consequence. This provides a performance estimate.
9. *Calculate a cost estimate.* Use technology-specific delivery dates to calculate each technology cost. Aggregate to a weapon-system cost. This provides a cost estimate.

10. *Obtain distributions.* Repeat steps 6 through 9 until the MC analysis is complete. Report schedule, performance, and cost distributions.

Throughout the report, we reference this methodology outline to provide orientation for the reader.

Organization of the Report

This report documents the risk-informed trade-space methodology and associated RTRAM tool jointly developed by the AMSAA Risk Team and the RAND team. Following this introductory chapter, Chapter Two discusses the incumbent risk methodologies, the organizations involved, the use in the acquisition process, and challenges to using the incumbent methodology. Chapter Three describes a theoretical framework that can be used for risk-adjusted trade analysis and documents its structure and how this framework might be used. Chapter Four describes the implementation of this framework in the context of Army weapon-system acquisitions and current AMSAA data-gathering methodologies, including the risk workshop used in support of AoAs. Chapter Five documents the mathematical structure of the conceptual model of Chapter Four, including model parameterization, structure, and outputs. It provides detailed mathematical formulations that will be relevant to technical readers interested in gaining a deeper understanding of the model structure. Chapter Six provides a notional example of the use of the model using risk workshop data and median cost point estimates for an AoA update of an example MDAP. Chapter Seven presents conclusions, including identified weaknesses in the approach and potential future work.

Four appendixes are included in the report. Appendix A provides a user manual for the delivered version of RTRAM currently coded in Microsoft Excel. Appendix B outlines the specific assumptions, methods, and functional forms used in the delivered version of the RTRAM software.[4] Appendix C documents best practices in expert elicitation (such as that used to elicit data during AMSAA's risk workshop). Finally, Appendix D provides the actual Visual Basic for Applications (VBA) code used in RTRAM as delivered.

---

[4]   We include this documentation in the appendix to emphasize the fact that the theoretical framework of Chapter Three and conceptual model of Chapter Four are quite general and can be adapted to different circumstances and user needs as necessary. We not only anticipate these changes but actively encourage users to adapt the model to their needs.

CHAPTER TWO

Current Cost, Schedule, and Performance Risk Methodologies Within the Army

The risk of a program going over budget (i.e., cost risk), going over schedule (i.e., schedule risk), and not performing to a set of originally designated physical characteristics (i.e., performance risk) are three major concerns for any acquisition program. The Army has existing methodologies for calculating such risks with the intention of trying to predict and mitigate them. Three Army organizations perform analyses to calculate different aspects of the cost, schedule, and performance risks of a program under the Army's consideration.

First, AMSAA, located at Aberdeen Proving Ground, Maryland, performs a technology-level risk assessment of cost, schedule, and performance for new Army systems (Bounds, 2014; Henry, 2012). It also performs a schedule risk assessment at the alternative level (Bounds, 2014). AMSAA is an Army Materiel Command organization that conducts these and other analyses to provide decision-relevant information to senior-level Army and DoD officials.

Secondly, ODASA-CE, located at Fort Belvoir, Virginia, performs system-level cost estimation analyses for programs that have not yet reached MS A. The Deputy Assistant Secretary of the Army for Cost and Economics (DASA-CE) is the principal adviser to the Assistant Secretary of the Army for Financial Management and Comptroller (ASA[FM&C]) on all Army cost and economic analysis activities.

Finally, TRAC, located in multiple locations, including White Sands Missile Range, New Mexico, and with headquarters at Fort Leavenworth, Kansas, performs system-level cost estimation and operational analyses for future Army systems. As an analysis agency of the

U.S. Army, TRAC conducts operations research on potential military operations worldwide to inform decisions about issues that the Army and DoD face.

All three organizations provide important input into the AoA process, and, although considerable efforts have been made to facilitate communication between them, the analyses they perform occur relatively independently of one another. Figure 2.1 depicts this relationship. Each oval represents an outcome produced (i.e., calculated or elicited) by the organizations described above, as well as others, such as program managers. Arrows with solid lines represent linkages that AMSAA typically includes in its analysis, while dashed lines represent contributions from other organizations. As shown, AMSAA's risk workshop effort provides the beginnings of a linked system in which to perform trade-space analysis. Notably, AMSAA has attempted to involve DASA-CE and TRAC, as well as others, in its technology-level risk assessment. The organization has noted in its risk guidebook (Bounds, 2014) that participation from the following organizations is desired: AMSAA; ODASA-CE; U.S. Army Training and Doctrine Command

Figure 2.1
Independence of Cost, Schedule, and Performance Risk Assessments

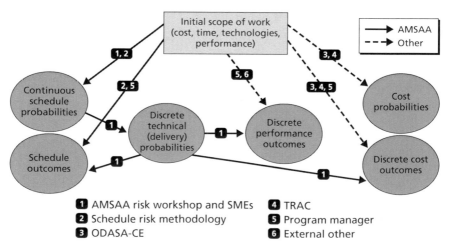

NOTE: Each oval represents an outcome produced (i.e., calculated or elicited) by the organizations described above, as well as others, such as program managers.
RAND RR701-2.1

(TRADOC) Centers of Excellence; U.S. Army Research, Development and Engineering Command (RDECOM); program executive officer or program manager (PM); Headquarters, Department of the Army (HQDA), and OSD action officers; TRAC; and the Army Capabilities Integration Center (ARCIC). For instance, AMSAA's analysis requires information about the current readiness levels (e.g., technology readiness level [TRL]) of each technology, which is provided by a Technology Readiness Assessment (TRA) prepared by the Deputy Assistant Secretary of the Army for Research and Technology for pre–MS B AoAs. Additionally, AMSAA's risk assessment relies heavily on values elicited during a risk workshop from SMEs representing many of the organizations previously mentioned.

The Current U.S. Army Materiel Systems Analysis Activity Methodology

AMSAA conducts a technology-level risk assessment of cost, schedule, and performance for new Army systems (i.e., technical risk assessment) and additionally perform a schedule risk assessment at the alternative level. We briefly describe each in this section. For a full description of the technical and schedule risk-assessment processes, see Bounds (2014) and Henry (2012).

Technical Risk Assessment

AMSAA's Technical Risk Assessment measures the risk that a technology relevant to an Army acquisition system will not be delivered (i.e., developed, integrated, and manufactured) within the desired time frame, cost target, and performance designation. The assessment is heavily informed by risk workshops that the organization conducts at the beginning of each program. The workshops, which include an elicitation of information from SMEs, determine the inputs that will eventually lead to a determination of the technology, cost, and schedule risk associated with technologies within a program. The AMSAA Risk Team or a designated contractor facilitates these workshops, in which as many SMEs as possible meet in one location to participate in the group exercise. SMEs may also participate via teleconference.

The risk workshops include several steps. To provide inputs to the workshop, the PM and RDECOM identify the KTs to be considered and their readiness levels (TRL, manufacturing readiness level [MRL], and integration readiness level [IRL]). These determinations are made based on Army TRA guidance (Pub. L. No. 111-23, 2009). This set of information is then provided to a set of SMEs participating in the risk workshop. Both technical and nontechnical SMEs are encouraged to participate in the risk workshops, and attendees may include participants from the PM office, Maneuver Center of Excellence, TRAC or DASA-CE, ARCIC, HQDA, OSD AoA action-officer stakeholders, the AMSAA Risk Team, RDECOM Systems Engineering Group, and technology SMEs.

During the risk workshop, SMEs are tasked with identifying risk drivers, estimating the time until KTs are at specific readiness levels, and assessing consequence levels if a technology is not delivered by the MS C date. First, SMEs are asked to identify any known technical risks or risk drivers associated with each KT that may affect schedule or cost risk, identify analogous programs for schedule and cost risk, and review the TRLs of each KT. Next, SMEs provide their judgment of the minimum, most-likely, and maximum times (in months) required for the technology to reach TRL 7 and the additional minimum, most-likely, and maximum times necessary to reach IRL 8 and MRL 8. Finally, the SMEs are asked to assess the consequences (i.e., cost, schedule, and performance) of each KT not reaching TRL 7, IRL 8, and MRL 8 by the proposed MS C date.[1] Consequence values are elicited on a scale from 1 (very low consequence) to 5 (very high consequence). Table 2.1 shows the definitions of those consequence levels.

With these risk workshop inputs, the technical risk assessment involves calculating the probabilities that a technology will reach

---

[1] TRL 7 is defined as "system prototype demonstrated in operational environment." IRL 8 is defined as "functionality of integrated items demonstrated in operational environment." MRL 8 is defined as "pilot line capability demonstrated; ready to begin low rate initial production." See Henry, 2012.

Table 2.1
Consequence Definitions Used in the U.S. Army Materiel Systems Analysis Activity's Risk Workshop

Level	Technical Performance	Schedule	Cost
1	Minimal consequences to technical performance but no overall impact [on] the program success	Negligible schedule slip	Pre–MS B: ≤5% increase from previous cost estimate Post–MS B: limited to ≤1% increase in . . . PAUC or . . . APUC
2	Minor reduction in technical performance or supportability [and] can be tolerated with little or no impact on program success	Schedule slip, but able to meet key dates (e.g., PDR, CDR, FRP, FOC) and has no significant impact [on] slack on critical path	Pre–MS B: >5% to 10% increase from previous cost estimate Post–MS B: ≤1% increase in PAUC [or] APUC with potential for further cost increase
3	Moderate shortfall in technical performance or supportability with limited impact on program success	Schedule slip that [affects] ability to meet key dates (e.g., PDR, CDR, FRP, FOC) [or] significantly decreases slack on critical path [or both]	Pre–MS B: >10% to 15% increase from previous cost estimate Post–MS B: >1% but < 5% increase in PAUC [or] APUC
4	Significant degradation in technical performance or major shortfall in supportability with moderate impact on program success	Will require change to program or project critical path	Pre–MS B: >15% to 20% increase from previous cost estimate Post–MS B: ≥5% but <10% increase in PAUC [or] APUC
5	Severe degradation in technical [or] supportability threshold performance; will jeopardize program success	Cannot meet key program or project milestones	Pre–MS B: >20% increase from previous cost estimate Post–MS B: ≥10% increase in PAUC [or] APUC danger zone for significant cost growth and Nunn-McCurdy breach

SOURCE: Henry, 2012.

NOTE: PAUC = program acquisition unit cost. APUC = average procurement unit cost. PDR = preliminary design review. CDR = critical design review. FRP = full-rate production. FOC = full operational capability. A Nunn-McCurdy breach occurs when an MDAP "experiences an increase of at least 15% [in PAUC or APUC] above the unit costs in the Acquisition Program Baseline" ("Nunn-McCurdy," 2002). It derives its name from Senator Sam Nunn and Representative Dave McCurdy because the amendment they proposed to the National Defense Authorization Act for Fiscal Year 1982 (Pub. L. No. 97-86, 1981) created the threshold.

TRL 7, IRL 8, and MRL 8 by a proposed MS C date using the minimum, most-likely, and maximum readiness-level times (i.e., triangular distributions). Because the probability of reaching each readiness level is considered independent from the others, the likelihood of the technology *not* reaching TRL 7, IRL 8, and MRL 8 is simply the product of the three probabilities subtracted from 1. This probability is then converted to an ordinal scale from 1, signifying less than 20-percent likelihood, to 5, signifying at least 80-percent likelihood, as shown in Table 2.2.

An overall consequence level for the technology is then calculated as the maximum of the three consequences elicited (i.e., cost, schedule, and performance). Finally, the final likelihood and maximum consequence are translated into a qualitative risk rating of low (green), medium (yellow), or high (red). This process is shown in Table 2.3. For a full description of the technical risk-assessment process, see the risk guidebook (Bounds, 2014).

Schedule Risk Assessment

AMSAA also developed a system-level schedule risk modeling approach called Schedule Risk Data Decision Methodology (SRDDM). The approach uses historical data in conjunction with SME-provided information to estimate the probability distribution related to system out-

Table 2.2
U.S. Army Materiel Systems Analysis Activity Likelihood-Level Definitions

Level	Likelihood	DoD Guidance (%)	Probability Range (%)
1	Not likely	~10	$L \leq 20$
2	Low likelihood	~30	$20 < L \leq 40$
3	Likely	~50	$40 < L \leq 60$
4	Highly likely	~70	$60 < L \leq 80$
5	Near certainty	~90	$L > 80$

SOURCE: Henry, 2012.
NOTE: L = likelihood.

Table 2.3
U.S. Army Materiel Systems Analysis Activity Technical Risk Assessment: Notional Example

Critical Technology	P(TRL 7)	P(MRL 8)	P(IRL 8)	L	Likelihood Level	C_P	C_S	C_C	Consequence Level	Risk Rating	Risk Driver
Transmit antenna	0.92	0.9	0.9	0.25	2	3	1	2	3		
Receive antenna	0.92	0.9	0.9	0.25	2	3	1	2	3		
Processor software	0.73	0.9	0.7	0.54	3	5	4	4	5		Software development not yet started
Processor electronics	0.96	0.9	0.9	0.22	2	2	1	2	2		
Barrel	0.98	0.9	0.9	0.21	2	2	1	1	2		
Receiver	0.87	0.9	0.9	0.30	2	4	3	1	4		Gun requires upgrade for required operational performance. Has not been demonstrated yet
Feeder	0.68	0.9	0.7	0.57	3	4	1	1	4		Gun requires upgrade for required operational performance. Has not been demonstrated yet

SOURCE: Bounds, 2014.

NOTE: P = probability of a given outcome. C_P = performance consequence. C_S = schedule consequence. C_C = cost consequence.

comes through MC simulations. These distributions are used to calculate the (discrete) probabilities associated with completion of acquisition phases. Outputs include the cumulative probability distributions associated with completion as a function of time for each phase. For a full description of SRDDM, see Bounds (2014).

Relationship Between Technical Risk Assessment and the Schedule Risk Data Decision Methodology

The technical risk-assessment methodology is focused on technology-level risks and uses SME opinion to generate distributions related to technology-specific schedules and categorical (i.e., integer-valued) system-level outcomes in the performance, schedule, and cost dimensions. SRDDM is focused on system-level acquisition-phase schedule risk. Despite their differing purposes, SRDDM and the technical risk-assessment methodologies are interdependent in that SRDDM uses some information gathered in the risk workshop as inputs and both provide information about potential schedule likelihoods and outcomes. As such, both provide useful information about schedule probabilities that can be used to analyze system-level acquisition risks in this dimension. However, a few key differences between the underlying data, calculation, and result reporting preclude these methodologies from becoming a truly linked assessment.

First, SRDDM uses data and includes analysis at the system level, while the technical risk assessment's data collection and analysis are performed at the technology level. The latter then aggregates technology-level results into a system-level risk determination. Although the technology risk assessment explicitly incorporates trades and risk mitigation to occur at the technology level, SRDDM provides a quantitative framework that the technology risk assessment lacks. Another strength of SRDDM is its ability to make explicit use of historical data. That is, although the SMEs' inputs into the technology risk assessment are informed by knowledge about analogous programs, SRDDM can use the schedules of analogous programs as input into its model. Overall, the two assessments both have strengths but lack a cohesive, holistic, quantitative framework that can be used to provide analysis related to schedule, cost, and performance trades of technologies and their associated risk-mitigation actions.

Challenges to Using the U.S. Army Materiel Systems Analysis Activity's Risk-Assessment Methodologies

As previously noted, AMSAA has taken initial steps to link cost, schedule, and performance risks using its risk workshop process. The workshop itself involves various acquisition stakeholders and requires communication among a broad set of SMEs. However, the main purpose of the workshop—to elicit SME judgments—also presents some limitations to the technical risk assessment. An SME elicitation is certainly necessary because the type of data AMSAA elicits does not generally exist elsewhere in an easily utilized form. Elicitations of any kind, though, are fraught with bias regardless of how well they are performed. This is because, as rational as any expert may try to be, the expert cannot control certain mental shortcuts that he or she will take when making quick judgments, for instance, about the time until a technology has reached TRL 7. Mental shortcuts, or heuristics, are used by anyone who needs to make quick judgments because, by definition, they ease the cognitive load of making a decision (e.g., Hastie and Dawes, 2010).

Heuristics, such as *availability,* or the tendency to overestimate the probability of events that are easy to recall (Hastie and Dawes, 2010), and *overconfidence,* or the tendency to underestimate the uncertainty surrounding certain elicited quantities (Morgan and Henrion, 1990), will certainly affect the results of AMSAA's risk workshop and the overall recommendations that can be made from the technical risk assessment. Therefore, the elicitation facilitator must take precaution when designing and implementing his or her associated protocol (i.e., the plan and script to follow for elicitation). There are established methods that can be used to ensure that the effects of heuristics during an expert elicitation are minimized (Hastie and Dawes, 2010). A further discussion of these heuristics, how they could manifest themselves during the risk workshop, and methods for minimizing their bias during elicitations can be found in Appendix C.

One other concern with using SME-elicited data is whether SMEs have the proper expertise to provide relatively accurate judgments of the quantities elicited. Cost estimation or consequence determination may be a particularly problematic area for SMEs. The overconfidence heuristic, for instance, may result in SMEs underestimating the likelihood of certain risky events (or not considering some set of scenarios) that could lead to cost growth. By not accounting for these risky events, SMEs could severely underestimate cost consequence. As such, we recommend that risk workshop moderators fully probe the SMEs to ensure that the elicited probability distributions are representative of all *possible* schedule outcomes, especially for events that might cause schedule outcomes different from those that are most likely (the mode of the distribution). In addition, users might consider using distributions with continuous support and soliciting information about, say, the 5th and 95th percentiles of schedule outcomes and fitting distributions of a different family (e.g., normal or lognormal) that do not have maxima or minima.

Representation of outcomes, or consequences, in the performance, cost, and schedule dimensions as elicited in the risk workshop is also a concern. Each consequence assumes an ordinal value between 1 and 5. These values are elicited from SMEs during the risk workshop as a relative value to a technology or system's baseline. Thus, a cost consequence for an engine of 3 may represent an increase in costs of 10 percent relative to some baseline cost of that engine. On the other hand, a cost consequence for a software system of 3 would also represent an increase in costs of 10 percent but relative to the baseline cost of that system. When comparing these two technologies, the 3 elicited for the engine represents a very different absolute cost from that for the software system. Because the objective of risk-informed trade-space analysis is ultimately to compare performance, schedule, and cost outcomes both within and between alternatives, this inconsistency in baseline technical performance levels has the potential to obscure key technological differences between systems. As such, we strongly recommend utilizing measures that are comparable across systems. Specific challenges with using this performance-consequence elicitation are discussed in Chapter Four. The methodology we developed, documented in subsequent chapters, does not rely on these consequence values for schedule or cost. However, performance-consequence values are still utilized in RTRAM.

A similar issue exists in the treatment of likelihoods in the technology risk assessment, which converts the continuous TRL, IRL, and MRL distributions (over time) to probabilities associated with a discrete outcome (e.g., technology readiness by MS C). This discrete probability is then the only information from that part of the elicitation that is carried on into the overall technical risk assessment. As discussed in later chapters, retaining the full probability distribution allows a decisionmaker to explore the range of possibilities for schedule, cost, and performance, rather than those at only one point in time (i.e., MS C), and can always be used to determine probabilities of being above or below a threshold (as in SRDDM).

Another challenge with the use of the risk workshop data in the technology risk assessment is that the data and the risk assessment link only the outcomes of performance, schedule, and cost through the discrete outcomes elicited from the SMEs. For instance, if one technology had a high likelihood of its schedule slipping, this would likely increase development costs. This is represented in the current methodology as a categorical consequence, rather than a quantitative estimate of the cost increase. A linked model would treat the acquisition process as a system of interconnected relationships that predicts *quantitative outcomes* in performance, schedule, and cost dimensions and allow users to investigate the implications of taking some sort of risk-mitigating action (allowing for performance or schedule slippage or cost overruns). This information is likely to be of considerable use to decisionmakers.

CHAPTER THREE

Theoretical Framework for Risk-Informed Trade-Space Analysis

What Is a Risk-Informed Trade Space?

Virtually all acquisition processes create outcomes in multiple dimensions. Some of these outcomes are goods, in the sense that more of this outcome is better. Examples may include certain performance parameters of a particular system, such as firepower, horsepower, or other positively valued characteristics. On the other hand, some outcomes may be bads, in that less of an outcome is preferred. Examples may include redefined performance parameters, such as horsepower shortfalls, but also include bads, such as months to a particular milestone and costs. In most, if not all, cases, there is unlikely to be a proposed system that simultaneously maximizes all of the good outcomes while minimizing all of the bad outcomes. As such, decisionmakers are forced to make trade-offs between these competing dimensions. This is recognized in the Weapon Systems Acquisition Reform Act, which requires that DoD "ensure that mechanisms are developed and implemented to require consideration of trade-offs among cost, schedule, and performance" (Pub. L. 111-23, Title II, § 201[a][1]). Here, the acquisition dimensions of interest are cost, schedule, and performance, though the exact definitions can be context-specific.

In this report, we define *trade space* as the set of multidimensional outputs or outcomes of interest from a particular process or set of choices (including the acquisition process and choices therein) that cannot be simultaneously optimized. In other words, the trade space maps from a set of defined alternatives, or a bundle of choices, to a set of outcomes in the multiple outcome dimensions of interest. The frontier (or Pareto frontier) of the trade space is the subset of outcomes corresponding to alternatives for which one cannot increase the outcome for a good or decrease the outcome for a bad by choosing another alternative without increasing or decreasing a good or bad value, respectively, in another dimension.[1] In other words, it is the set of nondominated alternatives, which is termed *technically efficient* in the economics literature. The optimal choice for the decisionmaker depends on preferences regarding the relative values of each dimension but will lie on the frontier. As such, a trade-space analysis does not identify a preferred alternative but rather provides decision support by offering information about the trade-offs between alternatives. Figure 3.1 is a graphical representation of a deterministic trade space.

A risk-informed trade space augments a deterministic trade space by recognizing that outcomes in one or more dimensions of interest are stochastic at the time of the analysis. As such, the outcomes are described probabilistically as a collection of weighted possible outcomes rather than as a single point in each dimension.

Although it is conceptually straightforward, a risk-informed trade space provides an abundance of additional information to a decisionmaker over which he or she must trade. For example, from the distribution in only one dimension, one could theoretically calculate any of the statistical moments of that distribution (e.g., mean and variance). Because potential decisionmakers have differing risk preferences, the frontier of a risk-informed trade space is not well-defined like the theoretical frontier shown in Figure 3.1. One decisionmaker might be willing to accept a smaller mean in a performance dimension for reduced variance in the cost dimension, and another may not be willing to make that trade. In other words, dominated alternatives (technical efficiency) are no longer easily definable because we are unable to state unambiguously which outcome distributions are preferred to others. Figure 3.2 displays selected information from a risk-informed trade

---

[1]   This term has a long history in both economics and system engineering. For a discussion in the context of the former, see Varian (1992) or Mas-Colell, Whinston, and Green (1995). Examples of the latter with a focus on uncertainty can be found in, e.g., Mattson and Messac (2005) and Daskilewicz et al. (2011).

Figure 3.1
A Two-Dimensional Deterministic Trade Space with
One Dominated Alternative

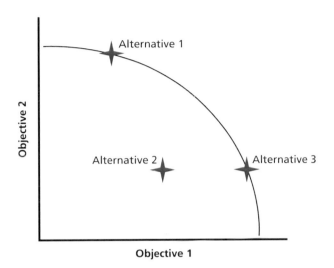

NOTE: Assume that objectives are good and that the analysis
has been restricted to three alternatives. Alternatives 1 and
3 are nondominated because no alternative ranks higher
in either objective 1 or 2. Alternative 2 is dominated by
alternative 3 because it ranks lower on objective 1 but the
same on objective 2. Without alternative 3, alternative 2
would be a point on the frontier.
RAND RR701-3.1

space in which objectives 1 and 2 are stochastic for each alternative,
markers correspond to the means of each alternative in each dimension, and error bars represent the 95-percent confidence intervals of the
outcomes for each alternative in each dimension.

Use of the Risk-Informed Trade-Space Framework

A risk-informed trade space allows for an investigation of the ex ante
(before the event) trade-offs between the stochastic outcomes of inter-

Figure 3.2
A Two-Dimensional Risk-Informed Trade Space

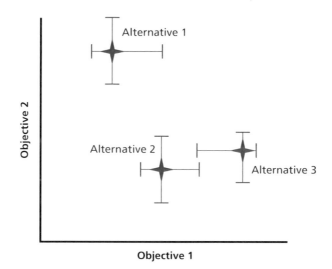

NOTE: Assume that objectives are good and that the
analysis has been restricted to three alternatives. Markers
indicate mean objective levels for each alternative. Error
bars provide the 95-percent confidence interval for each
alternative in each dimension.
RAND RR701-3.2

est.[2] It does so by estimating the outcomes of a partially controllable
uncertain process and using the distributions associated with those
outcomes to examine the implications of making changes to the process. Estimation of the risk-informed trade space is a modeling exercise
that produces outputs that can be used to identify and evaluate potential trades.

In the context of the weapon-acquisition process, the framework
is designed to answer a variety of questions, including, but not limited
to the following:

- What are the performance, schedule, and cost implications of
  delivering a specific, unproven technology?

---

[2] The event here in the acquisition context is the termination of the planning horizon by,
e.g., reaching a milestone or delivering a system.

- For a given materiel alternative, what are the performance, schedule, and cost implications if a specific technology is eliminated or traded?
- For a given materiel alternative, what are the performance, schedule, and cost implications if a schedule deadline (such as a project milestone) is relaxed or accelerated?
- For a given materiel alternative, what are the performance, schedule, and cost implications if research investment is curtailed or increased?

These questions imply that the outcomes of interest are related to the performance, schedule, and cost of the weapon system, which are uncertain at the time of analysis because of the nature of the overall research and development process. Partial control of the process is achieved through changing specific technologies, increasing or decreasing schedule constraints, or adjusting budgets.

A General Risk-Informed Trade-Space Methodology

The previous subsections defined a risk-informed trade space and argued that it could be used for decision support when faced with stochastic acquisition (and other) decisions. In this section, we describe a general methodology that can be used to estimate the outcome distributions. Figure 3.3 provides a visual interpretation of this process.

Design variables are choices related to the process being modeled. Essentially, they are the inputs into the trade-space process that the decisionmaker or designee could manipulate to change one or more outcomes of interest. They are the source of partial control of the outcomes. A unique set of design variables is termed an *alternative*.

For example, if one were modeling a financial portfolio–allocation process, the design variables might be the portion of the portfolio allocated to each of a countable number of stocks, bonds, and mutual funds. The decisionmaker has control over the elements in the portfolio, which will change the expected distribution of returns (an outcome of interest).

Figure 3.3
Estimating a Risk-Informed Trade Space

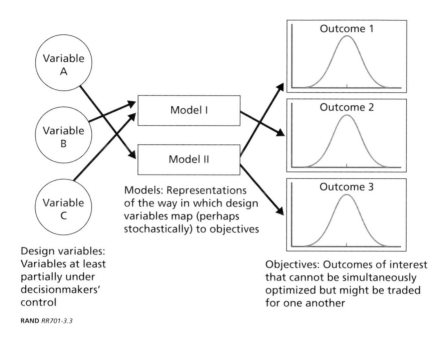

In the context of a pre–MS C AoA, the design variables might be the key technologies that make up a particular weapon system plus any potential COA that could be taken over the course of the process (such as allowing more development time or increasing funding levels). Changing any or all of these variables will change the distribution of the outcomes of interest, presumably organized around performance, schedule, and cost.

An *objective* is a measurable outcome of interest that is potentially tradable. That is, a decisionmaker might be willing to sacrifice some quantity of this outcome if he or she were able to obtain something else of value (more of a good, less of a bad). The mechanism for making a trade is through changing one or more of the design variables. If an objective is stochastic, as in a risk-informed trade space, then it is measured as a probability distribution over the potential feasible outcome levels. In the portfolio-allocation example, the objectives might be variables related to rates of return on the investment (for example, expected returns and the variance of those returns). In the AoA example, the objectives could be measures related to the cost, schedule, and performance dimensions of the weapon system under consideration.

A *constraint*, on the other hand, must be satisfied in order for an alternative to be viable. An example of a constraint set in the weapon-acquisition process is the KPPs that must be met.

Each alternative is mapped to the set of objectives via a set of one or more *models*. Models are formal or informal representations of the relationship between the design variables and the objectives, which take the design variables as inputs and create the objective distributions as outputs. Models need not have a one-to-one relationship with objectives; in fact, in many cases, a model might be used primarily to represent the interconnections between the design variables and the outcomes of interest. In the portfolio-allocation example, one might use a financial statistical model to estimate the risk and return of a particular asset. In the AoA case, the incumbent AMSAA methodologies can be considered one set of models. In subsequent sections, we present the newly developed RTRAM, which uses information from AMSAA's risk workshop and a host of other assumptions and data to estimate distributions related to performance, schedule, and cost in a theoretically consistent and interconnected manner.

CHAPTER FOUR

Proposed Conceptual Model of Risk-Informed Trade-Space Analysis in the Acquisition of Weapon Systems

In this chapter, we discuss the three major objectives or outcomes of interest that are relevant to trade-space analysis of major weapon systems and propose an approach to quantitatively and consistently estimate ex ante (prior to realization) output distributions related to performance, schedule, and cost based on AMSAA's risk workshop and previously developed methodologies.[1] We also provide recommendations for future data collection and parameterization.

Performance Outcomes

Information Available from the Risk Workshop

AMSAA's risk workshop provides information on SME-elicited opinions of the likelihood and consequence of failing to deliver each KT on time (hereafter, *nondelivery*) identified for each system considered in the analysis (Henry, 2012). In particular, the workshop output provides the likelihood and performance consequence of a particular technology not reaching TRL 7, IRL 8, and MRL 8 by the proposed MS C date, as well as identifying any known technical risks associated with the technology (Henry, 2012).[2] Performance consequences of nonde-

---

[1] *Consistently* here does not refer to the statistical property of consistency of an estimator; rather, we refer to a methodology that can be replicated in a similar manner across various contexts.

[2] Although our discussion focuses on MS C as the critical milestone date for concreteness, any potential milestone date could theoretically be considered.

livery are defined on a 1-to-5 integer scale, ranging from 1 (minimal consequences to technical performance but no overall impact to the program success) to 5 (severe degradation in technical or supportability threshold performance; will jeopardize program success).[3] See Chapter Two for more details.

The elicited consequence levels provide a KT-specific measure of the potential outcomes of technology nondelivery. These directly inform step 3 (determine consequence) in the methodology outlined in Chapter One.

Nature of Uncertainty

The source of uncertainty at the KT level is the discrete, stochastic event of nondelivery at MS C. In the event of delivery, the planned technology has reached the required TRL, IRL, and MRL, and the consequence level is presumed to be 0. In the event of nondelivery, the integer consequence level is realized.

Recommendations for the Future

At present, the performance consequence of technology nondelivery is assessed relative to a system's baseline, planned technology level at the time of elicitation. These baselines may differ across system alternatives and thus may have differing technical characteristics (such as speed or lethality). Because the objective of risk-informed trade-space analysis is ultimately to compare performance, schedule, and cost outcomes both within and between alternatives, this inconsistency in baseline technical performance levels has the potential to obscure key technological differences between systems. For example, weapon systems with different maximum ranges, and thus performance capabilities, would admit identical consequences of 0 if both were delivered. Thus, this metric does not allow for identification of the differing ranges. As such, a decisionmaker's ability to use this metric to answer questions about the relative costs (in the other dimensions) of achieving a particular technology level is diminished.

---

[3]   As explained in Chapter Two, AMSAA also elicits schedule and cost consequences along a similar 1-to-5 integer scale.

For this reason, we recommend that AMSAA develop and integrate baseline-independent technical performance metrics into the risk workshop process. These measures could be objective and relative to an appropriate measurement scale where available (e.g., range, speed, weight), objective and relative to a baseline technology (e.g., range over the incumbent technology, speed relative to one of the analyzed technologies), or subjective and relative to a baseline technology (e.g., a 1-to-5 scale of capability improvement relative to one of the analyzed technologies). The important attribute of such measures is that they can be used to compare alternative technologies relative to some standardized baseline.

Schedule Outcomes

Information Available from the Risk Workshop and Other Methodologies

AMSAA's risk workshop provides information on SME-elicited minimum, most-likely, and maximum times required to get each KT to TRL 7, as well as the same information for the technology achieving IRL 8 and MRL 8 conditional on reaching TRL 7 (Henry, 2012). These quantities can naturally be interpreted at triangular probability distributions that, when combined, define the subjective probability distributions of each potential schedule outcome (measured in, say, months to MS C) related to each technology. These directly inform step 2 (determine schedule) in the methodology outlined in Chapter One. As shown in Figure 4.1, AMSAA uses these triangular distributions to obtain a discrete probability of the likelihood that the technology will be delivered by MS C. AMSAA also solicits the schedule consequences of a technology-delivery failure on a 1-to-5 scale similar to that for the technical consequences.

In addition to these elicited distributions, AMSAA has developed SRDDM, which uses historical comparable program data at the system level, when sufficient, to estimate the probabilities associated with meeting planned first-unit-equipped dates for the systems under consideration (Henry, 2012). See Chapter Two for more details.

Figure 4.1
U.S. Army Materiel Systems Analysis Activity Methodology to Obtain a Probability of Delivering a Technology on Time from Elicited Triangular Distributions

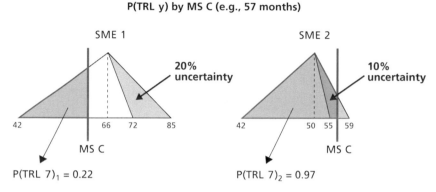

SOURCE: Bounds, 2014.
RAND RR701-4.1

Regardless of the method used, both techniques provide information that can be used to estimate the probability distribution that describes the potential schedule outcomes associated with a weapon system. SME opinions could also be used.[4]

Nature of Uncertainty

There are multiple driving sources of schedule uncertainty, including engineering hurdles (such as those related to development, integration, and manufacturing), testing delays, and contracting issues (Henry, 2012). The natural representation of schedule uncertainty is a probability distribution that describes every possible time to reach a particular program milestone or other date of interest within the range of defined possible outcomes.[5]

---

[4]   As described in Chapter Two and Appendix C, there are limitations regarding the use of SME opinions in this manner. Necessary cautions, such as methods to minimize heuristic processes, ensuring SME expertise, and proper documentation of the basis of SME judgment, should be undertaken.

[5]   This distribution can be continuous or discrete, depending on the level of fidelity needed by the user.

Recommendations for the Future

The implicit conceptualization implied by AMSAA's schedule consequence is that individual technology risk can drive overall system schedule outcomes. This is an inherently sensible proposition. However, for use in trade-space analysis, one would ideally like a continuous distribution that describes stochastic system-level schedule outcomes, rather than the discrete probability and consequence of not meeting a particular deadline for a particular technology. In this way, decision-makers can make decisions based on an evaluation of likely system outcomes, consistently with the framework presented in the previous chapter.

Moreover, as recognized by AMSAA's incumbent risk methodologies, schedule outcomes are driven by technology development and either drive or can be driven by cost outcomes. In other words, the three broad outcomes of interest are functionally interrelated, at both the technology and the system levels. Ideally, a fully formed trade-space analysis should recognize these interconnections, such that any proposed changes in design variables percolate through all outcomes in each dimension.

Finally, as previously mentioned in the section on performance outcomes, the 1-to-5 schedule-consequence values elicited may use different schedule baselines across system alternatives. For example, weapon systems with different schedule delivery times (SDTs) would admit identical consequences of 0 if both were delivered by their respective SDTs. However, the SDT for one weapon system may be much shorter than for another. Thus, this metric does not allow for identification of the differing schedule ranges.

Cost Outcomes

Information Available from the Risk Workshop and Other Methodologies

AMSAA's risk workshop solicits cost-consequence information for technology failures in a manner similar to that of technological and schedule consequences (1-to-5 integer scale; see Table 4.1). However,

Table 4.1
Example Cost Consequence and Level
Mapping

Level	Cost
1	≤5% increase from cost estimate
2	>5–10% increase from cost estimate
3	>10–15% increase from cost estimate
4	>15–20% increase from cost estimate
5	>20% increase from cost estimate

SOURCE: Henry, 2012.

this is the only cost information currently generated by AMSAA, and it has been noted that most technical experts do not have cost expertise (Younossi, Lorell, et al., 2008).

ODASA-CE and TRAC perform formal cost analysis for Army weapon systems. In general, these organizations use a formalized plan of work (such as a WBS) to describe the physical system being developed and procured and use various techniques (such as analysis of comparables) to parameterize the average or marginal costs of each component. Garvey (2000) describes the WBS as a "math model" of the cost of an engineering system; essentially, the system cost is represented as the sum of N WBS cost elements, each of which is a random variable with (perhaps) a functional relationship, a statistical relationship, or both. A detailed description of costing methodologies is beyond the scope of this project. Interested readers can refer to the *Department of the Army Cost Analysis Manual* (U.S. Army Cost and Economic Analysis Center, 2002).

To date, there has been little interaction between AMSAA, ODASA-CE, and TRAC on cost outcomes.

Nature of Uncertainty

There are multiple driving sources of cost uncertainty. For example, the *Department of the Army Cost Analysis Manual* (U.S. Army Cost and Economic Analysis Center, 2002), citing Garvey (1993), identifies requirement or configuration uncertainty, technical or system definition uncertainty, and cost estimation uncertainty. Later in that document (U.S. Army Cost and Economic Analysis Center, 2002), the manual identifies the following risk elements:

- performance-related risks: This category, related to requirement uncertainty, includes technical risk, configuration uncertainty, supportability risk, and programmatic risk. In general, these categories represent uncertainties or risks related to the development of the physical system or the program that controls it. Because most pre–MS C systems are not completely developed, there is no certainty regarding the exact specifications of the physical good that is to be ultimately fielded.
- schedule-related risks: This category relates to uncertainty in the realization date of a particular goal or set of goals. Interestingly, the *Department of the Army Cost Analysis Manual* (U.S. Army Cost and Economic Analysis Center, 2002) states, "schedule duration is affected by requirements and cost changes and for this reason, the schedule risks may be acerbated by the degree of requirements and cost estimating uncertainty" (p. 179).
- cost-estimating risks: This category relates to the uncertainties associated with the parameterization of cost elements within an estimate. These can occur for a variety of reasons but arise primarily because of information about parametric relationships that cannot be known at the time of analysis.

The *Department of the Army Cost Analysis Manual* (U.S. Army Cost and Economic Analysis Center, 2002) also states, "More often than not, cost estimating and schedule uncertainty are a reflection of technical, programmatic, and supportability risks" (p. 179). The approaches presented thereafter typically involve either adjusting a point estimate using SME or other information, performing sensitivity analysis on non–cost system parameters (e.g., weight), or performing sensitivity analysis or applying probability distributions to cost parameters. With respect to the latter, the manual notes the importance of recognizing whether the parameter uncertainty reflects only cost-estimating uncertainty or baseline (physical or schedule) uncertainties.

This documentation clearly shows that the U.S. Army is aware of the potential sources and drivers of cost risks, and the analyses of internal costing organizations, such as ODASA-CE and TRAC, reflect this recognition when they use probability distributions to represent cost uncertainties. The primary shortcoming, however, is that these analyses are not necessarily designed to document a clear causal relationship between underlying cost drivers and the resultant cost estimates. In other words, given an aggregate probability distribution that describes the likelihood of various cost levels, it is extremely difficult to uncover the performance and schedule outcomes associated with a particular cost outcome.

Furthermore, there is an implied, rather than explicit, link between behavior to mitigate risk and cost outcomes. For example, failure of a particular critical technology to be developed at a particular milestone might result in purposeful substitution of the incumbent (or another) technology, with delivery at the scheduled milestone but possibly a different cost structure. Similarly, if the technology truly is critical, then a decisionmaker might decide that additional time or investment (cost) is warranted. These events can be viewed as stochastic at the time of analysis (usually well before the milestone) and can affect the realized cost (as well as other dimension) outcomes. Although these potential actions might be incorporated into any distributional assumptions used in the cost-analysis methodologies, the lack of a specific functional relationship between actions and performance, schedule, and cost outcomes precludes answering questions about the trade-offs involved in taking such actions.[6]

Accurate reflection of these relationships would ideally be incorporated into a risk-informed trade-space analysis in order to be consistent with the technological and schedule dimensions. The natural

---

[6] This is not a criticism of those methodologies per se because they are not, in fact, designed to take this type of explicit relationship into account or to provide decision support over multidimensional outcomes. Rather, they are designed to estimate costs for a specific system, defined as a particular set of technologies developed and procured in a particular time frame, taking into account the relevant uncertainties where appropriate.

representation of cost uncertainty is an approximately continuous probability distribution that describes each possible system cost outcome and the associated probabilities.

Recommendations for the Future

As mentioned in the section on performance and schedule outcomes, the 1-to-5 cost-consequence values elicited may use different cost estimate baselines across system alternatives. For example, weapon systems with different cost estimates would admit identical consequences of 0 if both were delivered within this estimate. However, the cost for one weapon system may be much greater than for another. Thus, this metric does not allow for identification of the differing cost ranges.

Furthermore, because current Army costing procedures do not typically incorporate explicit structural relationships between performance, schedule, and cost estimates and the task of performing cost analysis typically falls to different organizations that perform technical and schedule analysis, there is a need to develop additional cost methodologies that are consistent with the technological and schedule risks described herein. In the following sections, we provide an initial framework for incorporating cost information into a risk-informed trade-space analysis. However, full development is beyond the scope of this research.

Courses of Action

Because the acquisition process is stochastic in nature, one cannot be certain of either the technology-specific or system-level outcomes before the process has completed. However, the Army can at least partially control the process by changing certain aspects of the system. We call such decisions *COAs*, and they must be specified as part of a system alternative in order to fully define it. These COAs directly inform step 4 (choose COAs) in the methodology outlined in Chapter One. Table 4.2 provides descriptions for the COAs.

Formally, COAs are decision rules that define what will be done in the event that a system is not fully developed (i.e., has not met the

Table 4.2
Courses of Action

COA	Description
COA Extend	Schedule is extended; technology can be delivered at some time after milestone date, time drawn stochastically from user-defined distribution
COA Finance	Additional financing; technology is forced to be delivered at or before milestone date with additional cost corresponding to stochastic draw from user-defined distribution
COA Replace	Technology replaced; the counterfactual technology (or same technology with worse performance) is delivered at milestone date if stochastic draw from user-defined distribution is after milestone date

TRL, IRL, and MRL) as specified at the milestone date. Choosing a COA may be necessary because of the possibility that the physical system cannot be delivered with the technologies specified within the original schedule and budget. In such a case, technology, schedule, or cost outcomes must necessarily be different from those originally assumed. The COAs can also be used as risk-mitigation tools because they are under the decisionmaker's control. We assume that COAs can be taken at the technology level.

The first COA, COA Extend, is to not enforce the original milestone date. This amounts to extending the schedule to allow for full development of a particular technology, rather than assuming that an alternative will be used for that system. The second COA, COA Finance, is to enforce the milestone date but provide additional financial resources in order to accelerate technology development. The third COA, COA Replace, is to enforce the milestone date and keep funding at the originally planned level, thus necessitating the replacement of the planned technology with some other solution or allowing the planned technology to be used with degraded performance. We term this replacement technology the *counterfactual technology*.

There is one COA associated with purposely changing a characteristic in each objective dimension (e.g., cost, schedule, performance) in order to keep the characteristic of the other two constant in the event of technology nondelivery. This provides the framework for analyzing decisions and trades on the basis of these dimensions.

A Proposed Linked System Architecture

We have argued in the previous sections that technical, schedule, and cost outcomes can be viewed as the relevant objectives for a risk-informed trade-space analysis that is consistent with the Weapon Systems Acquisition Reform Act of 2009, and we have discussed the information available for parameterizing a model that links the technical specification of a materiel system and the associated COAs to distributions that describe potential performance, schedule, and cost outcomes. In this section, we describe such a model in the context of Army weapon-system acquisition and the tools that AMSAA has previously developed for technical and schedule risk assessment. The model structure follows along with steps 6 through 10 in the methodology outline presented in Chapter One.

Figure 4.2 is a conceptualization of a structure that can be used to estimate the technical, schedule, and cost distributions that characterize the objectives of a single alternative within a risk-informed trade space. We discuss the model in general in this chapter (with reference to the associated steps from Chapter One's methodology outline); a detailed mathematical development is given in Chapter Five.

Use of the model begins with a specification of the KTs that make up the alternative of interest (step 1), the COAs to be taken if each technology fails to be delivered by the milestone (or other) date (step 4), the milestone date itself (step 2), and the objective metrics of interest. The user defines these. Associated with each KT is a counterfactual technology that would be used as a substitute if the KT failed to be delivered by the milestone date if the COA allowed for counterfactual technologies. Although it is helpful to conceptualize this counterfactual as a physical system, nondelivery of a KT manifests itself in the output of the model as a difference in the performance metric between the counterfactual and the KT (step 3). In other words, the specified KT has a performance consequence of 0, while its counterfactual has a performance consequence of KT nondelivery assigned during AMSAA's risk workshop.

Also associated with each KT is a user-specified schedule distribution for that technology, representing the (joint) probability that it will

Figure 4.2
Proposed Linked System Architecture

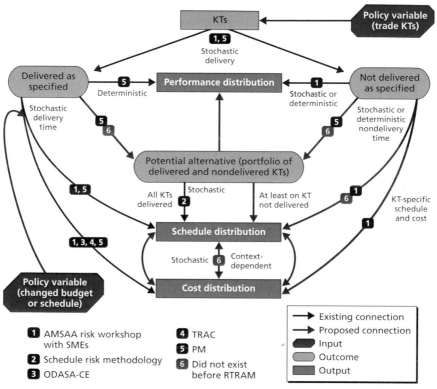

RAND RR701-4.2

achieve readiness levels of TRL 7, MRL 8, and IRL 8 at each future point in time. This distribution can be used to estimate the probability that a KT is at readiness levels less than TRL 7, MRL 8, and IRL 8 at any point in time (step 2). Then the likelihood of KT nondelivery is simply the probability associated with a schedule outcome later than the milestone date (assuming that the COA for that KT allows for counterfactual technologies).

Once the schedule distributions and their relationship are specified (e.g., independence between KTs), draws from the distributions are taken to represent one realization of the time to readiness levels of a future system (step 6).[7] This collection of dates is then translated into a future realized system on the basis of the specified COAs. If the milestone date is enforced according to the COA, then any schedule draw of a KT later than the milestone is deemed to be nondelivered. If the milestone date is not enforced (because the COA assumed is to either extend the schedule or increase the budget by a sufficient amount to induce on-time delivery), then the system deems the KT to be delivered. In the case of a COA that allows schedule slippage past the milestone date for one or more KTs, then the maximum of (1) the maximum time drawn (across such technologies) and (2) the minimum of the maximum time drawn for the other technologies and the milestone date is presumed to be the delivery date. KTs with schedule draws beyond the delivery date that allow for counterfactual technologies are considered not delivered. Following these rules, a future realized system is a portfolio of realized KTs and counterfactual technologies, with a particular performance profile as defined by the performance metrics associated with the realized KTs. In some cases, it may be necessary to aggregate the performance metrics to a system level (step 8).[8]

The performance metric at a system level depends on the portfolio of realized technologies that define the realized alternative. The schedule and cost estimates of a realized alternative depend on the schedule draws of each KT, the COAs assigned to each technology, and how the schedule draws are aggregated to a system level. Note that performance metrics for an individual KT are not stochastic with a realized alternative; that is, there is an assumed, deterministic relationship between a portfolio of KTs and the performance metrics. On the other hand, the

---

[7] Independence is not necessary but simplifies the process of sampling from the distributions. In many cases, however, independence may not be a realistic assumption. For example, when completion of a first KT is strictly required to complete a second KT, the schedule distribution of the second is dependent on the outcome of the first. These KTs are clearly not schedule independent.

[8] This is the case with the subjective performance-degradation measure captured by AMSAA's risk workshop. This measure incorporates both technology-specific and system-specific subjective impacts measured at the technology level; as such, aggregation is necessary in order to report a system-level measure of degradation when multiple technologies are not delivered.

realized date of technology readiness (and thus the associated costs) is stochastic within an alternative; that is, a single collection of realized KTs and counterfactual technologies admits its own probability distribution across schedule and cost dimensions.

To obtain one schedule and cost realization of the (now-defined) realized alternative, one must first define the specified delivery date of that system, which critically depends on the COAs chosen for the technologies. If any KT is allowed to develop past the milestone date, then the model determines the system's delivery date by the latest date drawn for each of these technologies (or the milestone date if all draws for these KTs are earlier than the milestone), as described earlier. If no schedule slippage is allowed for any KT, then the delivery date is assumed to be the earlier of the milestone date and the latest draw across all technologies. The delivery date is the realization of the schedule outcome (step 7).

Costs are jointly determined by the realized portfolio of KTs and the schedule draws for each individual KT, coupled with the user-specified parameters that determine fixed and variable (with respect to time) costs across each dimension (step 5). The fixed-cost parameters represent all non–time-varying, technology-specific costs for each KT and counterfactual technology. The variable-cost parameters represent the marginal costs of each KT with respect to time—that is, the costs incurred per unit time for each KT and counterfactual technology.[9] If schedule slippage is allowed and realized at the system level (i.e., the delivery-date value is greater than the milestone-date value, meaning that the delivery date is later than the milestone date), then total costs are computed as the sum of (1) fixed plus variable costs for all delivered KTs through the delivery date, (2) the sum of fixed and variable costs for all nondelivered KTs through the milestone date, and (3) the sum of fixed and variable costs for all counterfactual KTs from the milestone date to the delivery date.

---

[9]   Through the parameterization of the model, the user ultimately determines the planning horizon and, thus, the interpretation of fixed and variable costs. For example, if life-cycle costs were being modeled, the user could decide to treat all time-varying costs after the delivery date as fixed.

If schedule slippage is not allowed and the chosen milestone date is the binding constraint at the system level, then total costs are computed as the sum of (1) fixed plus variable costs for all delivered KTs through the technology-specific schedule date, (2) the sum of fixed and variable costs for all nondelivered KTs through the milestone date, and (3) fixed costs for all counterfactual KTs.[10] In this situation, if a KT is delivered via a COA that adds funding to achieve the milestone date, then the cost associated with the drawn schedule overage for that KT is assumed to be the increased funding amount. In other words, the increased cost is determined by the model and internally consistent with the treatment of schedule. The calculated cost estimate is one possible realization of the cost outcome for the realized alternative (step 9).

The process described above essentially samples from the implied performance, schedule, and cost outcomes by making draws from the user-supplied distributions and estimating outcomes on the basis of these draws and the parameter values. Repeating this process a large number of times by taking independent draws, recording the results, and tabulating frequencies provides empirical estimates of the implied outcome distributions and is known as MC analysis (step 10). These empirical distributions can be used to tabulate any statistics of interest across the outcome distributions. Figure 4.3 provides a flow chart diagraming the logic of the model. Chapter Five details the mathematic specification of the linked system model.

_____

[10] This assumption can be relaxed to allow for differential variable costs during the development period if the user is willing to specify some rule about when the switch in technologies is made.

Figure 4.3
Internal Logic of the Linked System Model for One
Realization

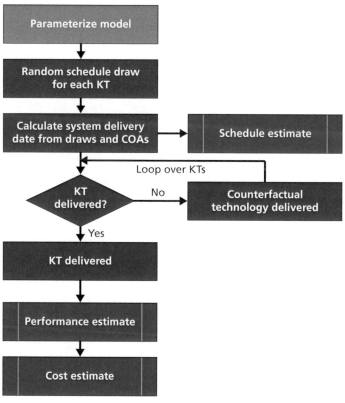

RAND *RR701-4.3*

CHAPTER FIVE

The Model

The RAND team developed RTRAM based on our interactions with AMSAA and other acquisition analysis teams and the framework detailed in the previous chapter. In this chapter, we detail the mathematical framework of RTRAM. It provides detailed mathematical formulations that will be relevant to technical readers interested in gaining a deeper understanding of the model structure. For reference, Table 5.1 presents a glossary of the variables, subscripts, and superscripts presented in this chapter. Additionally, the steps from the methodology outline in Chapter One are referenced for additional orientation. The mathematical framework is presented in the most general form possible to allow for future relaxations of the specific functional form and other assumptions used in the delivered tool. These assumptions and functional forms used for the tool delivered to AMSAA in September 2013 are detailed in Appendix B.

Table 5.1
Glossary of Variables, Subscripts, and Superscripts Referenced in Chapter Five

Variable	Description
π	Discrete probability of delivery or nondelivery
a	Realized alternative
c	Cost
coa	COA
fc	Fixed cost

Table 5.1—Continued

Variable	Description
g	Probability distribution, elicited from SMEs, associated with a KT achieving TRL 7, MRL 8, and IRL 8
kt	KT
ms	Milestone date
p	Performance metric
rt	Delivery date
t	Time
vc	Variable cost

Subscripts

k	KT
n	System or alternative
$n(e)$	Set of technologies associated with schedule extension
$n(-e)$	Set of technologies with COA other than schedule extension

Superscripts

$+$	Delivery of a technology
$-$	Nondelivery of a technology
a	Alternative
e	COA Extend (when used in conjunction with variable coa)
f	COA Finance (when used in conjunction with variable coa)
i	Delivery or nondelivery ($+$ or $-$)
K	KT
IRL	Time until IRL (when used in conjunction with variable t)
MRL	Time until TRL (when used in conjunction with variable t)
r	COA Replace (when used in conjunction with variable coa)
TRL	Time until MRL (when used in conjunction with variable t)

Model Parameterization

Consider system n, composed of $K > 0$ KTs (step 1). KT k can be delivered as specified within a system, denoted kt_{nk}^{+}, or not delivered as specified, with the system utilizing the assumed counterfactual technology, denoted kt_{nk}^{-}. Each $kt_{nk}^{i_k}, i_k \in (+,-)$ is associated with the following:

- a user-defined performance metric set (step 3)
- the information necessary to estimate the (unconditional) probability distribution associated with a KT achieving TRL 7, MRL 8, and IRL 8 elicited from the SMEs, denoted $g_{nk}(t)$ (step 2)
- the milestone date for the program, denoted ms (step 2)
- the fixed- and variable-cost parameters to be used in the analysis (step 5)
- the COA associated with each technology, denoted coa_{nk} (step 4).

Key Technology–Specific Schedule Distributions

The (unconditional) probability distribution associated with a KT achieving TRL 7, MRL 8, and IRL 8 at each point in time t elicited from the SMEs either directly or indirectly is denoted $g_{nk}(t)$.

Current practice in the risk workshop is to elicit three probability distributions directly from participants, assuming triangular distributions: (1) the unconditional distribution describing time to TRL 7, denoted $g_{nk}^{TRL}(t)$; (2) the distribution describing additional time to MRL 8 conditional on achieving TRL 7, denoted $g_{nk}^{MRL}(t)$; and (3) the distribution describing additional time to IRL 8 conditional on achieving TRL 7, denoted $g_{nk}^{IRL}(t)$. These must be aggregated to represent $g_{nk}(t)$. We describe one method of doing so in Appendix B.

The aggregation of the individual distributions associated with technology, manufacturing, and integration is a key step in the modeling process. It describes the possible schedule outcomes (step 2 of the methodology outline), which subsequently drive the performance and cost outcomes. Misspecification of the schedule probability distribution will result in misleading model outcomes. As a result, the user should be vigilant in ensuring that the unconditional schedule distribution encompasses all known interrelationships in the development process.

System Delivery Date

Define the milestone (or other planned schedule target) specified by the user as $ms > 0$, which is a constant denoting units of time in the future relative to the current, or starting, time period 0 (step 2). Possible COAs for each technology include (1) coa^e, which allows for schedule extension past ms to complete development of a KT; (2) coa^f, which adds financing (cost) to ensure timely delivery of a KT at ms; and (3) coa^r, which assumes replacing the KT with the counterfactual technology at ms (step 4).[1] Variants of these COAs are possible; for example, a user may want to estimate the consequences of allowing only a limited, rather than unlimited, schedule slippage. The mathematical implications of these variants can be derived from the discussion below and are not documented in this report.

Denote coa_{nk} as the COA that the user assumes to be associated with the kth KT for the nth system, and t_{nk} be the random variable described by $g_{nk}(t)$. If $coa_{nk} \neq coa^e \forall k$, then the delivery date, or realized time of the nth alternative, is determined by the earliest of (1) the date of completion of all KTs or (2) the milestone date ms. Formally, the delivery date $rt = \min\left(\max\left(t_{n1},\ldots,t_{nK}\right), ms\right)$. If, however, $coa_{nk} = coa^e$ for at least one k, and we denote the set of these technologies as $\mathbf{t}_{n(e)}$, then the delivery date

$$rt_n = \max\left\{\max\left(\mathbf{t}_{n(e)}\right), \min\left(\max\left(\mathbf{t}_{n(-e)}\right), ms\right)\right\},$$

---

[1] Given the model's structure, the counterfactual technology is simply a representation of the performance characteristics that the system assumes to be delivered at the schedule threshold. Depending on the circumstances, it could be interpreted as an incumbent, partially developed, or other technology.

where $\mathbf{t}_{n(-e)}$ is the set of technologies with COAs not equal to coa^e. Because the delivery date is a function of the random variables $(t_{n1}, \ldots, t_{nK})$, it is also a random variable.

Schedule Distribution

The schedule distribution of alternative n is the distribution of rt_n. It is derived from $g_{nk}(t)$, coa_{nk}, and possibly ms and describes the probability that a system will be delivered at a particular date t. As such, the distribution depends critically on the joint distribution of the random variables $(t_{n1}, \ldots, t_{nK})$, which the user must directly or indirectly specify. The easiest case is to assume independence, so that

$$g_n\left(t_{n1}, \ldots, t_{nK}\right) = \prod_k g_{nk}\left(t_{nk}\right),$$

though this assumption can be relaxed.

The distribution of rt_n can then be found analytically by the method of distribution functions or estimated numerically.[2] Numeric estimation simply involves randomly sampling from the joint distribution $g_n(t_{n1}, \ldots, t_{nK})$ and calculating a value of rt_n from the appropriate formula based on the coa_{nk}. Binned frequencies of the resulting rt_n values provide an estimate of the schedule distribution of system n. This provides the schedule estimate for step 7 of the methodology outline.

---

[2] The method of distribution functions is one of three major methods to calculate the distribution of a function of one or more random variables. It allows one to calculate the distribution of a single-valued function of multiple random variables x_n, say, $Y = f(x_1, \ldots, x_N)$. We denote this distribution as $g(y)$. It involves finding all values of $(x_1, \ldots, x_N)$ such that $Y = y$ and the region where $Y < y$, then calculating the cumulative distribution function where $Y < y$ by integrating the joint distribution of the x's, $g(x_1, \ldots, x_N)$. Other methods for calculating this distribution include via direct transformation or using the moment-generating functions of the random variables.

Realized System

Because the development time of each KT and the delivery date are random, there is a probability distribution that describes the possible realized systems (a portfolio of KTs and counterfactual technologies) at each system delivery date. This distribution is used in conjunction with the performance and cost parameters to estimate the performance and cost distributions associated with an alternative.

We begin by taking the system delivery date rt_n as given and calculated as described in the previous subsection. The discrete probability of delivery at rt_n for each KT, $\pi^+_{nk}(rt_n)$, is defined as

$$\pi^+_{nk}(rt_n) = \int_0^{rt_n} g_{nk}(t)\, dt.$$

The associated probability of nondelivery, $\pi^-_{nk}(rt_n)$, is defined by $\pi^-_{nk}(rt_n) = (1 - \pi^+_{nk}(rt_n))$ and corresponds to the probability of the counterfactual technology being delivered at rt_n.

A realized alternative is defined as a random variable $A^I_n = a(kt^{i_1}_{n1}, kt^{i_2}_{n2}, \ldots, kt^{i_K}_{nK}) = a(kt^i_n)$, where the superscript I corresponds to a unique identifier based on the portfolio of realized technologies for alternative n.[3] Given the discrete nature of delivery of each KT, there are thus 2^K potential outcomes $a(kt^i_n)$ of each alternative n. The probability of realizing each alternative is $\Pr(A_n = a) = g^a_n(kt^i_n)$, with $0 \le g^a_n(a(kt^i_n)) \le 1$ and

$$\sum_a g^a_n(a(kt^i_n)) = 1.$$

The function $g^a_n(a(kt^i_n))$ is implied through user-specified assumptions about the relationship between given technologies and the laws of probability. Assuming independence, for example, implies that $g^a_n(a(kt^i_n)) = \prod_k \pi^{i_k}_{nk}(rt_n)$. In this case, the probability of delivering

---

[3]   Note that the function $a(\ )$ is identical for each process and thus needs no additional identifying indices.

system n is simply the product of the probabilities that each KT can be delivered at rt_n. Conditional distributions could also be used if known or assumed (for example, in the case of a technology dependent upon completion of another). The support of this probability distribution is made up of the 2^K possible realized alternatives $a\left(\mathbf{kt}_n^i\right)$.

Performance Distribution

Each technology $kt_{nk}^{i_k}$ is assumed to have a deterministic relationship with performance outcomes.[4] As such, the performance of each realized alternative A_n^I is deterministic as well. Ex ante variation in performance is thus driven by the stochasticity of the realized system outcomes, which are, in turn, driven by the distributions of the individual KTs, the means by which they are aggregated into a system, and the (potentially endogenous) system delivery date.[5]

Formally, define a univariate performance metric associated with $kt_{nk}^{i_k}$ as $p_{nk}\left(kt_{nk}^{i_k}\right)$.[6] Current practice from the risk workshop defines

$$p_{nk}\left(kt_{nk}^{+}\right) = 0$$

for all n, which implicitly assumes that performance for delivery of any realized system made up solely of KTs, or $A_n^1 = a\left(kt_{n1}^{+}, kt_{n2}^{+}, \ldots, kt_{nK}^{+}\right)$, is equal.[7] From a pre-rt_n perspective, performance of the delivered system is a random variable through the stochastic delivery outcomes described by $g_n^a\left(a\left(\mathbf{kt}_n^i\right)\right)$. For an individual technology, the random

---

[4]   In the case of the risk workshop, this is the performance consequence that is realized if schedule targets are missed. The assumption of deterministic outcomes can easily be relaxed.

[5]   The term *endogenous* here refers to the fact that schedule outcomes are determined within the model.

[6]   The assumption of a univariate metric can be relaxed but has not been operationalized in the current version of RTRAM.

[7]   We strongly recommend reconsideration of this specification to enable meaningful between-system performance or technical comparisons.

variable describing the potential performance-consequence outcomes is thus denoted $P_{nk} = p_{nk}(kt_{nk})$.[8]

Because there are 2^K potential outcomes of each system n corresponding to delivery or nondelivery of each technology at the system delivery date, the performance distribution of alternative n is defined as a function of the random variables associated with each KT. Define the functional map from each of the 2^K potential outcomes to a performance measure as $P_n = p_n\left(p_{n1}(kt_{n1}),\ldots,p_{nK}(kt_{nK})\right)$.[9] The variable P_n is thus the performance measure of a realized system alternative on the basis of the technologies that appear in the system. For example, suppose one is modeling a system with two KTs. Assume that performance consequence is measured on a scale of 0 to 5, with 0 indicating delivery of the technology on time and as originally designed. The range 1 through 5 is the performance consequence of a technology not being delivered as originally designed, with 5 indicating the most negative consequence. Assume that, for technology 1, the consequence is 2 and that, for technology 2, the consequence is 4 (perhaps as solicited by SMEs). Further assume that the consequence to the system (as opposed to individual technology) is $p_n(\cdot) = \max(\cdot)$. In this case, if both technologies are delivered, then the system-level performance consequence is $P_n = \max(0,0) = 0$. If technology 1 is delivered but technology 2 is not, then the system-level performance consequence is $P_n = \max(0,4) = 4$. If, on the other hand, technology 1 is not delivered but technology 2 is,

---

[8]   In a case in which the performance consequence of a KT is simply measured as a scalar, then the resultant KT consequence distribution is identical to the delivery distribution except for the unit of measure. For example, if the probability of delivery for KT A is 0.85, with corresponding consequence levels of 0 for delivery and 2 for nondelivery, then the probability of consequence 0 is 0.85 and the probability of consequence 2 is 0.15.

[9]   This function must be specified and is not necessarily obvious. In RTRAM, consistently with current practice, we use the maximum of the individual consequences—that is, $Pn = \max\left(p_{n1}(kt_{n1}),\ldots,p_{nK}(kt_{nK})\right)$. In other cases, one might be interested in a performance index, which could be a weighted average of several performance metrics. In still other cases, one might allow for multiple performance dimensions. The user should take care that the performance measures used are (1) meaningful to the decisionmaker and (2) standardized such that between-system comparisons are meaningful. In other words, systems with identical performance metrics should have identical technical specifications.

then $P_n = \max(2,0) = 2$. Finally, if neither technology is delivered, then $P_n = \max(2,4) = 4$.

As with the schedule distribution, the performance distribution conditional on rt can then be found analytically by the method of distribution functions or estimated numerically. Conditional on an rt value, numeric estimation involves randomly sampling from the joint distribution, $g_n(t_{n1}, \ldots, t_{nK})$, comparing the drawn values for each technology to rt and constructing the realized alternative $a\left(\mathbf{kt}_n^i\right)$ accordingly, and calculating $P_n = p_n\left(p_{n1}\left(kt_{n1}\right),\ldots,p_{nK}\left(kt_{nK}\right)\right)$. Binned frequencies of the resulting P_n values provide an estimate of the performance distribution of alternative n conditional on rt. This provides the performance estimate for step 8 of the methodology outline.

Cost Distribution

We assume that cost is a function of time and of the KTs used to define an alternative (i.e., a function of the outcomes from the two other model dimensions [step 5]). The former are related to variable costs, while the latter are related to fixed (technology-specific) costs. Cost parameters can come from a variety of sources, including ODASA-CE or TRAC, the risk workshop, schedule risk methodologies, transformations thereof, or other outside sources. A functional relationship between time and dollars is necessary, even if it is a per-unit time calculation (linear specification). Depending on the data available, a simple approximation for the relationship between time and cost might be reached by taking a total cost estimate, subtracting any identifiable fixed costs, and dividing by the schedule presumed in the estimate to obtain the marginal and average costs per unit time. To the extent that data on schedule and cost from comparable programs or any KT-specific cost information are available, they could be used in estimating as well. Regardless of the data used to parameterize the costs, it is more important to give careful consideration to relative, rather than absolute, relationships.

In its most general form, the cost function for a realized system is represented by $C_n^I = c_n^I\left(t_{n1}^{i_1}, t_{n2}^{i_2}, \ldots, t_{nK}^{i_K}, kt_{n1}^{i_1}, kt_{n2}^{i_2}, \ldots, kt_{nK}^{i_K}\right)$, where the first K terms relate to the variable costs associated with randomness in technology-specific development and delivery times and the second K terms are associated with technology-specific fixed costs. Each term is a random variable, so costs are random as well. The current RTRAM assumes linearity in the cost function, which we detail below.

Denote the variable-cost parameters for each technology as $vc_{nk}^{i_1} > 0$ and the associated fixed-cost parameters as $fc_{nk}^{i_1} \geq 0$. [10] The development costs for each realized technology depend on the delivery status of the technology and the COA, coa_{nk}, associated with that technology. If the technology is delivered and $coa_{nk}^{i_k} = coa^r$ or $coa_{nk}^{i_k} = coa^e$, then the technology-specific cost function is $c_{nk}^{+} = fc_{nk}^{+} + vc_{nk}^{+} \times t_{nk}^{+}$. If the technology is delivered and $coa_{nk}^{i_k} = coa^r$, then the technology-specific cost function is also $c_{nk}^{+} = fc_{nk}^{+} + vc_{nk}^{+} \times t_{nk}$ because it is assumed that the extra investment necessary to complete the development process is equal to what would be required if the schedule were extended. If the technology is not delivered, implying that $coa_{nk}^{i_k} = coa^r$, then $c_{nk}^{-} = fc_{nk}^{+} + vc_{nk}^{+} \times rt_n + fc_{nk}^{-}$. Note that, under standard COAs, the variable costs for counterfactual technologies are assumed equal to those of the original technology; however, custom COAs may relax that assumption. [11] In addition, the fixed costs of the planned technology are always realized. Total system costs are estimated as

$$C_n^I = \sum_k c_{nk}^I.$$

As with the previous schedule and performance distributions, the cost distribution can be found analytically by the method of distribution functions or estimated numerically. To do the latter, the random sample from the joint distribution $gn(t_{n1}, \ldots, t_{nK})$ that determines the realized alternative $a\left(kt_n^i\right)$ is used to calculate the technology-specific

---

[10] Although the current version of RTRAM assumes that these parameters are not random, the conceptualization of the fixed and variable cost parameters as random because of uncertainty is conceptually straightforward.

[11] The logic is that variable costs are driven primarily by labor costs, which are likely driven by time. Custom COAs could relax this assumption, but the user must specify the time at which the counterfactual variable costs would begin to accrue.

and total costs of the realized alternative. Binned frequencies of the resulting C_n^I values provide an estimate of the cost distribution of alternative n. This provides the cost estimate for step 9 of the methodology outline.

CHAPTER SIX

An Example Application of the Risk-Informed Trade Analysis Model Using a Demonstration Tool

In this chapter, we present an example application of the model using the Excel-based demonstration tool RTRAM. This tool was designed to be an accessible interface for users to both input relevant data and view tailored output graphs. The example application illustrates RTRAM's capabilities in providing decision support for investigating trade-offs related to performance, schedule, and cost outcomes for an example MDAP. The case study is not intended to be a comprehensive analysis of the program; rather, it is a notional representation of the risk-informed trade-space methodology and associated RTRAM.

The first section of this chapter briefly describes the RTRAM input and output interfaces, while the second section provides the MDAP case study. For more-detailed documentation of these features, see Appendix A. The following sections report the findings of the original risk workshops and AoA and then illustrate how inputs from the risk workshop and AoA could be used in RTRAM.

The Risk-Informed Trade Analysis Model Interface

We developed RTRAM in Microsoft Excel to provide an accessible interface for users that allows for the input of relevant data and viewing tailored output graphs following the MC exercise. RTRAM uses VBA code and ActiveX controls, which allow it to be accessible to any user with access to Excel (i.e., no need for special software) and to be reasonably automated and user-friendly. Furthermore, the user is encouraged to adapt the code to his or her own needs (for example, by changing the specific functional forms documented in Appendix B or creating custom COAs).

RTRAM includes three major interface tabs: one for the user to make input choices and two that allow the user to view output from the model in a variety of tailored formats. One output tab allows for a more exploratory analysis of the possible alternative trades, while the other provides a higher-level comparison of a few alternatives. In addition, there are three tabs used for populating alternative names, KT names, and different schedule duration times. Appendix A provides a user manual for the tool. The remainder of this section provides a high-level description of the tool.

After a user has input the underlying parameters (e.g., alternative names, duration times, technologies), including any risk workshop, cost, and other data, RTRAM's input interface may be populated. Figure 6.1 provides a screenshot of the input interface that has been populated for an unspecified example. One the left side of the screen,

Figure 6.1
The Risk-Informed Trade Analysis Model Input User Interface

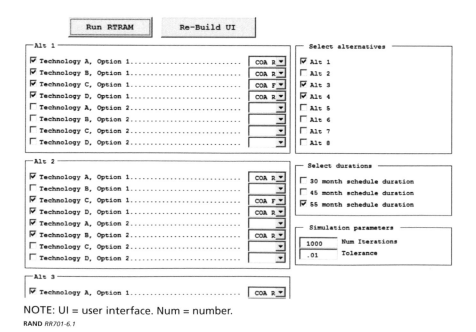

NOTE: UI = user interface. Num = number.

RAND RR701-6.1

the user can "design" each alternative. Alternatives are named in the upper left corner of each box (e.g., **Alt 1**, **Alt 2**). Each box then contains a list of all available technologies. A user may check the technologies included within that alternative. To the right of the technology name, the user also chooses the COA using the drop-down menu. Thus, the user designs the alternative by choosing a combination of technologies and technology-level COAs.[1] Once all alternatives have been defined, the user then may choose a subset of schedule duration times and alternatives to be run through the model. These choice boxes are shown on the right side of the interface. Choosing a subset will significantly reduce the computing time of the model. On a different tab, the user selects alternative names.

Figure 6.2 provides a screenshot of the first output interface: the risk graph. This interface provides numerous views of the schedule, cost, and performance (probability) distributions as a function of their consequence (e.g., months, dollars, and 0-to-5 degradation scale, respectively). Although schedule and cost distributions are presented as line graphs representing cumulative distributions (i.e., S-curves), the performance distribution is shown as a bar graph representing a probability density function. The bar/line distinction was chosen because the performance data are discrete, while the cost and schedule data are continuous. Furthermore, AMSAA decisionmaking surrounding the performance of alternatives was quite different from that for cost and schedule. AMSAA noted a preference for viewing each level of performance degradation separately, while the decision framing for schedule and cost (e.g., likelihood that an alternative will be delivered prior to the milestone date) was best answered by using cumulative distributions. These considerations were taken into account when choosing the

---

[1] A KT is associated with a certain fixed- and variable-cost parameterization; if the user wishes to use different parameterizations for different alternatives, he or she must define multiple KTs.

COA 2 extends schedule, COA 3 adds cost, and COA 4 allows for counterfactual technologies. They are equivalent to coa^e, coa^f, and coa^c in the text and were the originally developed names.

Figure 6.2
The Risk-Informed Trade Analysis Model Risk-Graph Interface

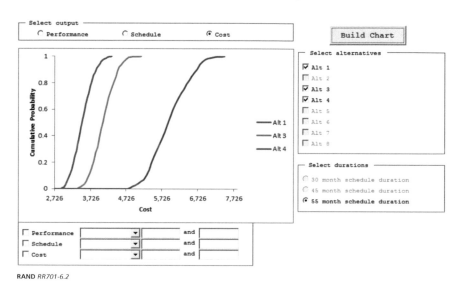

RAND *RR701-6.2*

current presentation.[2] This output interface was designed to enable the user to explore the entire decision trade space in great detail, allowing for the elimination of nonpreferred alternatives.

Across the top of the screen, the user may choose to view distributions for schedule, cost, or performance. The distributions of one or more alternatives are displayed as cumulative distribution functions. The user may choose this schedule duration time on the right side of the interface. Each schedule duration time is a particular assumed milestone date. The user also chooses the alternatives to be shown on the graph. In Figure 6.2, cumulative cost distributions are being shown for three alternatives (alternatives 1, 3, and 4) for a 55-month schedule duration.

In addition to this basic functionality, the user can tailor the risk graph to a variety of constraints. The user may choose to constrain cost, schedule, or degraded performance to be greater or less than any value or between any two values. For example, the user may be inter-

---

[2] The user may customize the displays by altering RTRAM's code because the underlying data allow for maximum flexibility.

ested to explore these schedule distributions if costs were constrained to below \$3.5 billion. The risk graph would then present the conditional cost distribution.

Figure 6.3 provides a screenshot of the second output interface: the three-dimensional graph. This interface provides a three-dimensional frontier of the mean schedule, mean cost, and mean performance values for each alternative under consideration. This output interface presents a higher-level comparison that may be most useful after choosing a set of preferred alternatives with the interface presented in Figure 6.2.

With performance on the x-axis and cost on the y-axis, each alternative's mean cost and performance are plotted on a scatterplot, with the alternative's mean schedule displayed as a data-point label. In Figure 6.3, for instance, the third and fourth alternatives (shown in green and red, respectively) have a mean schedule of 30 months, while alternative 1 (shown in blue) has a mean schedule of 58.5 months. On the right side of the interface, the user chooses the alternatives and the schedule duration, in months, to be shown on the graph.

With RTRAM and its underlying methodology, and given the proper underlying data, a decisionmaker may thoroughly explore the cost, schedule, and performance trade space of any weapon system during its acquisition process. The tool enables the user to develop

Figure 6.3
Cost Versus Performance Versus Schedule

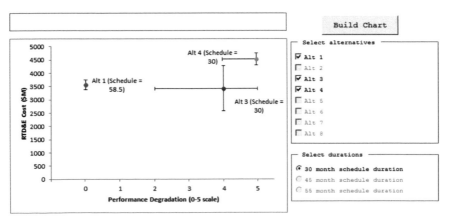

a reproducible and quantitatively based narrative surrounding risk-informed trades of multiple alternatives to help inform the judgments of acquisition decisionmakers.

Example Major Defense Acquisition Program Case Study

This section presents a case study using a previous AoA update to illustrate RTRAM and the demonstration tool's ability to display risk-informed trades of alternatives of an example MDAP. The Defense Acquisition Executive directed the Army to develop requirements for a newly conceptualized MDAP, terminating an old similar program. The weapon system that was being used at the time had significant vulnerabilities, was overburdened as a result of increased requirements, and did not have the capacity for new, desired equipment. CAPE approved a variety of alternatives for the MS B AoA: four systems, as well as a base case:

- base case: This is the currently fielded weapon system.
- upgrade: This alternative would include developing and procuring an upgraded version of the weapon system that included all upgrades that were possible without actually modifying the design.
- upgrade + redesign: This alternative would also include developing and procuring an upgraded version of the current weapon system, but, unlike the upgrade alternative, it would require a redesign. It would include all of the modifications of the upgrade alternative. In addition, however, the size of a major system component would be increased.
- new concept: This alternative was a new-concept design of the weapon system that met all requirements set out in the concept design document (CDD).
- concept stripped: This alternative is a stripped version of the new-concept alternative. It is CDD compliant but has reduced weight when compared to the new-concept alternative.

To inform Army decisionmakers about the choice of system, an AoA update for the MDAP was subsequently performed. As part of this AoA, ODASA-CE and TRAC performed cost analyses of the systems. Complementary to this AoA were the technical and schedule risk assessments that AMSAA performed for the four systems. To obtain data for the technical risk assessment, AMSAA held a risk workshop in which it elicited SME opinions about the weapon-system technology readiness, schedule, and costs. Table 6.1 shows a summary of key outputs from the analyses for three of the systems and their KTs considered in the AoA. The third and fourth columns provide a summary of the SMEs' assessments of risk of the technology and system, respectively. The fifth column provides the cost analysis from ODASA-CE and TRAC, and the final column provides the probability of meeting the MS C date as calculated as part of AMSAA's schedule risk assessment. Note that, although it was projected to have considerably higher RDT&E costs and to have a similar probability for meeting the proposed schedule, SMEs in the risk workshop rated the new-concept design as less risky than two existing systems.

As part of its technical and schedule risk assessments, AMSAA also performed a qualitative trade analysis using the data presented in the third and fourth columns of Table 6.1. Through a variety of behavioral actions (e.g., COAs), each system's technical risk could be partially mitigated. The impacts of these actions, however, are not limited to the performance dimension; rather, these trades also potentially affect schedule and cost outcomes. Although this trade analysis does provide a sense of some behavioral actions that could be performed to mitigate risk, the model that expressed linkages between cost, schedule, and performance was informal, largely subjective, and qualitative in nature. Even if the quantitative costs and schedule calculations shown in Table 6.1 were used, these quantities were assessed at the system level and, therefore, would not provide a direct mapping to technology-related trades. Overall, although the data and analysis performed by AMSAA, ODASA-CE, and TRAC provided the building blocks to inform trades between technologies and their schedule, cost, and performance outcomes, the information was not synthesized in a linked model. A linked model could, for instance, illustrate how the

Table 6.1
Technology, Schedule, and Cost-Analysis Results from Analysis-of-Alternatives Workshops

Alternative	Technology	Technology Risk (Likelihood, Consequence)	System-Level Risk	Life-Cycle Cost (billions of dollars)	RDT&E Cost (billions of dollars)	Probability of Completing EMD Phase and Passing MS C (%)
Upgrade	1	(5, 4)	High	32.6	1.0	67
	2a	(1, 3)				
	3	(5, 3)				
Upgrade + redesign	2	(1, 3)	High	50.6	2.4	67
	3	(5, 3)				
	4	(5, 5)				
	5	(1, 5)				
New concept	2b	(1, 3)	Moderate	62.5	7.3	70
	6	(1, 2)				
	7	(2, 5)				
	8	(1, 5)				

NOTE: EMD = engineering and manufacturing development.

decision to fund one technology early affected other technology- and system-level outcomes, or it could present the distribution of possibilities for schedule, cost, and performance for any technology or system.

Example Major Defense Acquisition Program Analysis Using the Risk-Informed Trade Analysis Model

RTRAM provides a quantitative means for linking the example MDAP data presented previously in this chapter in such a manner that it can inform decisions about the cost, schedule, and performance risks at the technology and system levels across the different COAs. In the following sections, we explore how results from a beta version of RTRAM, when loaded with these example MDAP data,[3] can provide a rich set of information for informing decisions about the alternatives.

Model Inputs

The primary data used in the analysis are the technology-specific TRL, IRL, and MRL triangular distributions elicited from the risk workshop, as well as the integer performance consequences of employing an implied counterfactual technology (see Table 2.1 in Chapter Two). A delivered technology is assumed to have a performance consequence of 0. The milestone date was assumed to be 65 months from the time of analysis.

Cost information was taken from the median point estimates of system-level RDT&E costs provided in the MDAP's MS B dynamic-update AoA. In order to demonstrate one potential process for using limited information to parameterize RTRAM, no additional cost information was used. Furthermore, using RDT&E costs was chosen over life-cycle costs because they more appropriately fit the scope of the research question. That is, because RTRAM used schedule data relevant only to the acquisition phases up to MS C, costs expended after

---

[3] In this analysis, we considered only the upgrade, upgrade-with-redesign, and new-concept alternatives. The figures presented here have been modified from the default Excel charts using alternative formatting options common to all Excel spreadsheets.

this time were not within scope and would therefore be treated as a constant. RTRAM focuses only on costs that are variable pre–MS C.[4]

In the analysis, it was assumed that RDT&E costs were 100 percent variable because there was no information about fixed costs. To decompose these estimates into technology-specific variable-cost parameters, we first attempted to reconcile the cost estimates and the SMEs' schedule estimates. To do so, we first estimated cost shares by assuming that each KT's share of development costs was equal to the maximum of the most likely time it was expected to take to complete (from the technology-specific distribution that describes schedule outcomes as computed from the individual SME TRL, IRL, and MRL distributions) or the assumed milestone date (65 months). We then used these weights to obtain technology-specific RDT&E costs and divided by total estimated system time to obtain the marginal variable-cost number. These calculations help demonstrate how RTRAM can be used with very limited information regarding expected costs.[5]

Formally, if μ_k is the mode from the technology-specific schedule distribution and ms is the milestone date (which is this application's schedule duration), then the share of RDT&E costs is

$$w_k = \frac{\max\{ms, \mu_k\}}{\sum_k \max\{ms, \mu_k\}}.$$

This formula essentially returns the share of overall expected system time dedicated to the development of a particular technology. Expected time is represented by the maximum of the milestone date or the most likely time to completion, as given by the numerator.[6] The denominator

---

[4]   Life-cycle costs could be calculated and would cause a shift in the cost distributions by a constant.

[5]   Of course, for accuracy, the user should include as much detailed information as is practical.

[6]   In this way, the information from SMEs and the contractor is included in the formula. If the SMEs deem that the technology will not be ready for delivery at the milestone date, then we defer to their judgment using this specification. If the SMEs deem that the technology

sums over all technologies. Letting *RDTE* denote total RDT&E costs for the system, the marginal (and, in this case, average) cost parameters are then computed as

$$mvc_k = \frac{w_k \times RDTE}{\max\{ms, \mu_k\}} = \frac{RDTE}{\sum_k \max\{ms, \mu_k\}}.$$

Marginal variable costs are thus the total variable RDT&E costs attributable to each KT divided by expected time to completion. Note that using this specification implies that marginal variable costs are constant across technologies (i.e., the right side of the previous equation does not depend on the *k*th technology except through the sum in the denominator) and depend on the total expected development time (if SMEs believe that the schedule is likely to slip) or the milestone date.

Because we assume zero fixed costs, this implies that technology-specific costs associated with this exercise are of the form $c_k = mvc_k \times t_k$, where t_k is the completion time for the technology.[7] Note that total technology-specific costs will depend on the drawn schedule outcome. As use of the model progresses, we anticipate that more-appropriate cost information can be developed for use in the model.

Finally, the user can control the number of MC iterations used in simulation. In what follows, we used 5,000 iterations.

Comparison of the Three Alternatives Allowing for Late Delivery

If we were willing to allow for late delivery of the MDAP systems under consideration, would any one system stand out as the best choice? RTRAM produces cumulative distributions for cost, risk, and schedule outcomes. From Figure 6.4, we see that, because we are allowing for late delivery, each system is given adequate time to reach a performance degradation of 0 (i.e., the desired performance). That is, each system is

---

will be ready before the milestone date, then we assume that some of the resources available in development will be shifted to the other technologies. In this way, draws from the left side of the schedule distribution do not imply considerable cost savings. Note that this is one of many possible specifications for introducing cost information into the model.

[7]   The completion time depends on the draws for all of the technologies and their COAs.

Figure 6.4
Performance-Degradation Distribution of Upgrade, Upgrade-with-Redesign, and New-Concept Alternatives for COA Extend

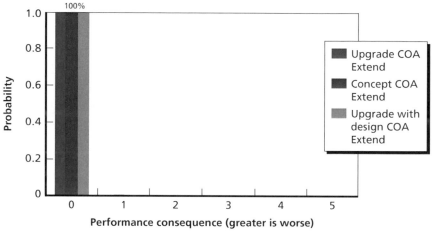

RAND RR701-6.4

100 percent likely to meet the desired performance. However, as shown in Figure 6.5, only the new-concept design (red line) will likely meet

Figure 6.5
Schedule Distribution of Upgrade, Upgrade-with-Redesign, and New-Concept Alternatives for COA Extend

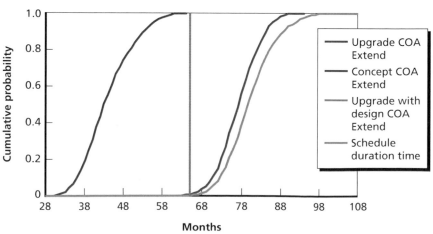

RAND RR701-6.5

the schedule duration. Both the upgrade and upgrade-with-redesign alternatives have very low likelihoods (almost 0) of meeting SDT.

When these alternatives' cost distributions are considered (Figure 6.6), we see that the new-concept alternative (with a range of $5.9 billion to $9.9 billion) will be drastically more expensive than the upgrade alternative ($0.9 billion to $1.3 billion). Additionally, the range of costs that the new-concept alternative could take on is much greater—the width of the 95-percent confidence interval for the cost of the upgrade alternative is around $0.3 billion, compared with $2.0 billion for the new-concept alternative. Thus, there is a cost–schedule trade-off between the upgrade and new-concept alternatives. The upgrade alternative is the least expensive option but with a longer schedule, whereas the new-concept design is the most expensive option but with a shorter schedule.

By comparing the three alternatives using COA Extend, we find that no alternative truly dominates (i.e., is best in all three dimensions) in this scenario. A decisionmaker would need to choose whether staying on schedule is worth more than $5 billion. Of course, the perfor-

Figure 6.6
Research, Development, Test, and Evaluation Cost Distribution of the Upgrade, Upgrade-with-Redesign, and New-Concept Alternatives for COA Extend

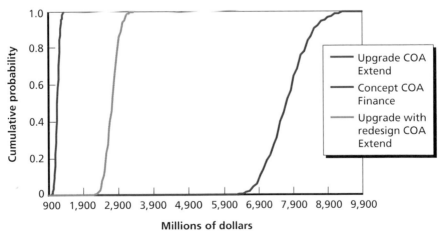

mance of these two systems could be very different, and this could heavily influence the decision. However, as discussed previously, the current framing of the performance consequence does not allow for this comparison. A final insight that may be gained from this assessment is that the upgrade-with-redesign alternative will likely be delivered the latest but will be the second most costly without intervention. This suggests that the other two alternatives dominate this one, and there is little reason to spend resources considering it in the absence of additional risk-mitigation actions.

Single Alternative with Different Courses of Action

If we were interested in a single system of this example MDAP, would any one risk-mitigation action stand out as the best choice? Here, we present this analysis for the upgrade alternative in which, for each alternative, all KTs within the alternative adopt the same COA. Figures 6.7 through 6.9 present the performance, schedule, and cost distributions, respectively. Although extending the schedule or financing for on-time delivery allows the system to remain at the desired performance level,

Figure 6.7
Performance-Degradation Distribution of the Upgrade Alternative Across the Courses of Action

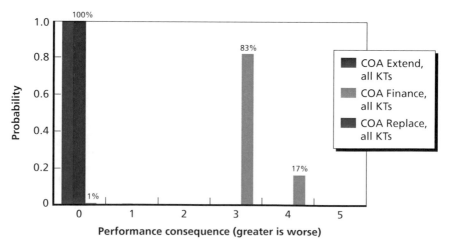

NOTE: Percentages might not sum to 100 because of rounding.
RAND RR701-6.7

Figure 6.8
Schedule Distribution of the Upgrade Alternative Across the Courses of
Action

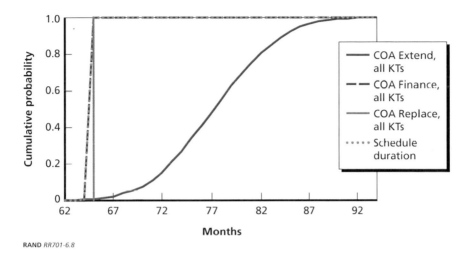

RAND RR701-6.8

Figure 6.9
Research, Development, Test, and Evaluation Cost Distribution of the
Upgrade Alternative Across the Courses of Action

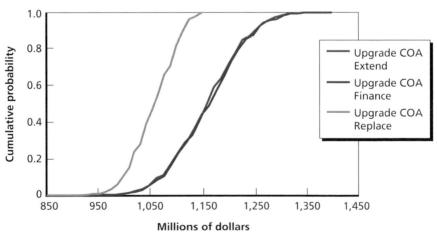

RAND RR701-6.9

there is only a 1-percent chance of avoiding significant performance degradation if the milestone date is enforced (Figure 6.7). However, accepting replacement technologies will allow for delivery by the SDT (Figure 6.8) and keep costs relatively low (Figure 6.9).

Comparing across the COAs, we see that there is a cost associated with meeting SDT at no consequence to performance (COA Finance) for the upgrade alternative relative to COA Replace (see Figure 6.9).[8] A decisionmaker would need to assess whether meeting SDT with no performance degradation is worth an increase of 10.4 percent in cost ($1.05 billion to $1.16 billion at the median). This may be conceptualized as a cost–performance trade-off, in which the lower-cost, worse-performing alternative (upgrade with COA Replace) may be compared with the most expensive but better-performing option (upgrade with COA Finance).

In addition, Table 6.2 presents discrete probability information about the performance, schedule, and cost distributions in a user-customized table for the upgrade alternative across the different COAs. The user chooses a set of constraints (right side of the table) by inputting them into the Excel table. With this information, RTRAM calculates specific probabilities based on those constraints. Here, we see that the likelihood of the upgrade alternative being delivered by a schedule duration of 65 months is 100 percent if COA Finance or COA Replace is used but less than 1 percent if COA Extend is implemented. Accordingly, the likelihood increases slightly for COA Extend of the upgrade alternative being delivered in 68 months or less, to 3.7 percent. Furthermore, there is a much higher likelihood (97.9 percent versus 37.0 to 37.7 percent) of delivering the upgrade alternative within a budget

---

[8]   We note that this cost differential is due to the treatment of variable costs in RTRAM. In order to deliver a program KT (i.e., kt^i) on schedule under COA Finance, the cost equivalent of delivery at a (late) draw is added to the costs associated with delivery at the schedule duration. As such, the cost distributions are identical when all kts are associated with either COA Finance or COA Extend. However, although the total RDT&E costs and technological specifications are identical, the timing of delivery differs (see Figure 6.8). The cost of enduring delivery at the schedule duration is calculated as the difference between COA Finance and COA Replace because the latter will always be the minimum-cost option if there are no fixed costs. Note that this analysis does not incorporate increased marginal variable costs (as opposed to total variable costs) for COA Finance.

Table 6.2
User-Customized Table Presenting Discrete Probability Information for the Upgrade Alternative with User Inputs and Three Courses of Action

Probability	COA Extend	COA Finance	COA Replace
Percentage by which the schedule will be less than the schedule duration: Pr($sched \leq duration$)	0.7	100.0	100.0
Percentage by which the schedule will be constrained: Pr($schedconstraint$)	3.7	100.0	100.0
Percentage by which cost will be constrained: Pr($costconstraint$)	37.7	37.0	97.9
Percentage by which performance will be constrained: Pr($perfconstraint$)	100.0	100.0	0.8
Number of months by which the alternative has a 50% probability of exceeding the schedule: Pr($sched \leq time$) =50%[a]	77	64	64
Millions of dollars by which the alternative has a 50% probability of being under budget: Pr($cost \leq budget$) = 50%[a]	1,153	1,160	1,053

NOTE: Constraints imposed are as follows: schedule cannot exceed 68 months, cost cannot exceed $1,137 million, performance consequence cannot exceed 2, table probability value (entered by the user for the last two rows of the table) is 50 percent, and schedule duration is 65 months.

[a] Values are approximate.

of $1,137 million if COA Replace is used. However, the probability that the upgrade alternative will have good performance is very low (0.8 percent). The table further provides information on the 50th percentile (user-chosen value) of the schedule and cost distributions. Thus, we see that 50 percent of the time, COA Finance and COA Replace will deliver the upgrade alternative within 64 months but that the 50th percentile of the cost distribution for COA Replace is the most favorable, at $1.05 billion. Overall, the table may provide the decisionmaker with the view that, if little importance is placed on the performance of the upgrade alternative, it would be best to implement COA Replace but that, if performance is important, COA Finance dominates COA Extend on every dimension except one.

Technology-Specific Courses of Action

Exploring the COAs for the upgrade alternative introduces questions about the drivers of the performance degradation, shown in Figure 6.7, and how to mitigate the risk associated with that degradation. The risk workshop data reveal that performance-degradation consequences for technologies 1 and 3 were driving performance for the upgrade alternative. From Figure 6.7, we see that technology 3 (performance consequence of 3) drives 83 percent of the performance risk, while technology 1 (performance consequence of 4) drives 17 percent of this risk.

RTRAM allows for the upgrade alternative to be explored when risk mitigation is directed at one of these technologies. Figure 6.10 shows the cost distributions for the upgrade alternative when all KTs receive COA Finance (red line), when all KTs receive COA Replace (green), and when all KTs receive COA Replace except technology 1, which receives COA Finance (blue line). Figure 6.11 shows a similar graph, except that the blue line represents when all KTs receive COA Replace except technology 3, which receives COA Finance. Figure 6.10 illustrates that mitigating performance risk for technology 1 only slightly increases the right tail of the cost distribution (blue line). That

Figure 6.10
Research, Development, Test, and Evaluation Cost Distribution, Technology 1 Mitigation of the Upgrade Alternative

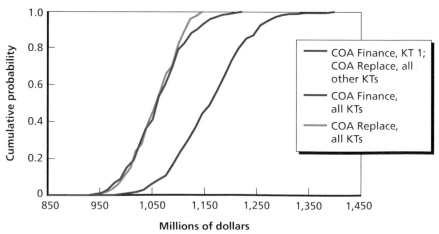

Figure 6.11
Research, Development, Test, and Evaluation Cost Distribution,
Technology 3 Mitigation of the Upgrade Alternative

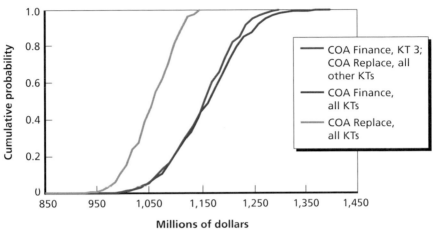

RAND *RR701-6.11*

is, a small increase in costs is only slightly more likely. On the other hand, Figure 6.11 illustrates that the same risk mitigation for technology 3 shifts the entire distribution, meaning that a larger increase in costs is much more certain. A comparison of mitigating the risk of each of these technologies suggests that it is considerably more expensive to perform risk mitigation for technology 3, which drives 83 percent of performance risk, than it is for technology 1, driving 17 percent of risk.

Comparison of Multiple Systems Across Different Courses of Action

If we were interested in two different systems of the example MDAP, would one risk-mitigation action (i.e., COA) stand out as being more or less effective for one of the systems than for another? We explore this question in Figures 6.12 and 6.13. Figure 6.12 illustrates that financing a project to meet the SDT (COA Finance, the blue and purple bars) will mitigate all of the performance risk for both the upgrade and upgrade-with-redesign alternatives. By definition, this mitigation will increase costs. Figure 6.13 shows, though, that this mitigation will cost between $130 million and $772 million for upgrade with redesign (green and purple lines), compared with only $15 million to

Figure 6.12
Performance-Degradation Distribution of the Upgrade and Upgrade-with-Redesign Alternatives for Various Courses of Action

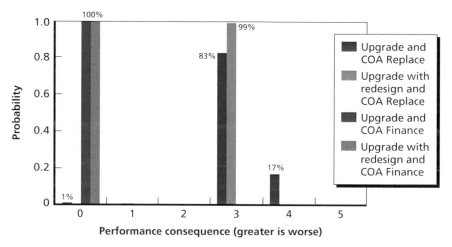

NOTE: Percentages might not sum to 100 because of rounding.
RAND *RR701-6.12*

Figure 6.13
Research, Development, Test, and Evaluation Cost Distribution of the Upgrade and Upgrade-with-Redesign Alternatives for Various Courses of Action

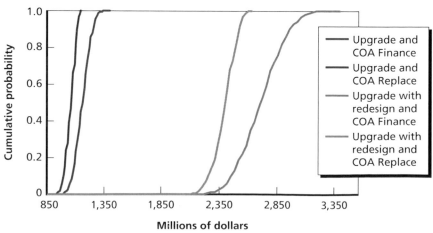

RAND *RR701-6.13*

$252 million for upgrade (blue and red lines). Furthermore, financing the upgrade alternative provides the same performance distribution as financing the upgrade with redesign but has much lower absolute cost. Therefore, RTRAM results suggest that upgrade with redesign is dominated by the upgrade alternative and that, as suggested earlier, it may not be appropriate to consider upgrade with redesign any further.

Three-Dimensional Analysis

Finally, the demonstration tool provides an overall comparison of the example MDAP alternatives across the three dimensions of cost, schedule, and performance. Figure 6.14 plots four alternatives against their mean performance degradation (x-axis), mean RDT&E costs (y-axis), and mean schedule (point labels). Ninety-five–percent confidence intervals are reported for cost and performance as well. To simplify the display and not cause information overload, we chose to not present confidence intervals for the schedule distribution. However, this information is available from RTRAM's output, and a simple pro-

Figure 6.14
Performance-Degradation Distribution Versus Research, Development, Test, and Evaluation Cost Distribution

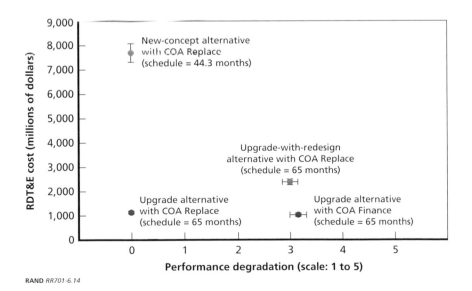

gramming change could allow for schedule confidence intervals to be presented next to the mean values in the point labels. Overall, we see that any system can provide a performance of 0, but not without some increase in cost or schedule. Although the new-concept alternative with COA Replace and the upgrade alternative with COA Finance have performance-degradation values of 0, the new-concept alternative is five times more costly than any other alternative, and the schedule for the upgrade alternative is much longer than that for the new-concept alternative. COA Replace alternatives for the upgrade and upgrade-with-redesign alternatives also have considerably longer schedules than those for the new-concept alternative. Therefore, no alternative is the dominant option (i.e., performs best on all three dimensions). Decisionmakers must decide whether their priorities lie in mitigating cost growth or schedule growth; they must decide whether better performance is worth the cost or schedule growth. The example MDAP case study illustrates RTRAM's ability to make these types of trades explicit.

CHAPTER SEVEN

Conclusions

The AMSAA Risk Team and the RAND team jointly developed a decision-support methodology for risk-informed trade-space analysis in weapon-system acquisition and the first iteration of RTRAM. The framework and model allow the user to investigate multidimensional trade-offs both within and between weapon systems prior to production using elements of system engineering, production economics, and risk analysis to functionally and probabilistically relate performance, schedule, and cost outcomes and their related uncertainties holistically and understandably. In the model, the technology-development process is conceptualized as a physical system consisting of a portfolio of technologies with associated technical capabilities, and the completion of each technology is stochastic (i.e., a discrete random variable). As such, the performance characteristics of the final system are stochastic. In addition, the time of technology development is also stochastic and, in part, drives the overall cost of the system.

A novel feature of the model is the incorporation of technology-specific COAs, or risk-mitigation behaviors, which take place in the event that the technology is not developed by the milestone date (e.g., allowing for performance degradation, schedule slippage, or increased investment). This allows for a quantitative evaluation of potential risk-mitigating actions across the multidimensional output space. The framework should be useful to those who wish to compare a set of alternatives in an acquisition environment, such as that completed in an AoA, using joint performance, schedule, and cost outcomes and uncertainties.

Model Limitations

As with all models, RTRAM has certain limitations, makes a few very simplifying assumptions, and can be improved in future work. In this section, we describe what we see as those issues, of which a user of the model or information from it should be acutely aware:

- The performance outcome lacks a meaningful baseline for between-system comparisons. As documented in Chapter Four, the performance outcome described by the risk workshop output is relative to a KT that is specific to a proposed weapon system, rather than to a reference technology or some other objective measure. In practice, this precludes meaningful between-system technical comparisons (though within-system comparisons are acceptable). We recommend that future users normalize performance metrics.
- The risk workshop data are based on subjective judgments. As described in Appendix C, subjective judgments from SMEs are subject to heuristic processes. Those data cannot be validated, nor have experts been calibrated to ensure some stability or realism in their opinions. However, to the extent that such stability or realism exists, distributions derived from historical data or other sources could be used as inputs into the model.
- Current functionality allows for only triangular TRL, IRL, and MRL input distributions. The theoretical model in Chapter Five allows for any properly defined technology-specific schedule distribution; however, the delivered version of RTRAM is hard-coded for the triangular distributions elicited from the risk workshop. In practice, this restricts the schedule distributions to have finite support (i.e., a hard minimum and maximum schedule), which may not be realistic. It would be a straightforward exercise to allow for alternative distributional families in the model.
- Parametric certainty is assumed. Although the framework accounts for linkages between technological, schedule, and cost outcomes, the parameters that govern these linkages are assumed known and fixed. For example, the performance consequence is

assumed to be a deterministic one-to-one map with a counterfactual technology. However, it is very possible that the impact of using a particular counterfactual may not be known with certainty, the counterfactual technology itself may not be known with certainty, or both. This would result in that map being stochastic rather than deterministic. Similarly, the fixed- and marginal or average variable-cost parameters are assumed known, but a variety of nonschedule factors can cause deviations from assumed budgets (see Bolten et al., 2008). Omission of these factors tends to underestimate the variance of the resultant outcome distributions, as well as skew them (depending on the assumptions). It would be a reasonably straightforward change to allow for stochastic parameters in the model.

- Cost data are limited. As documented in Chapter One, the WBS for budgeting typically does not include schedule as an input, and the curse of dimensionality (the exponentially increasing number of possible systems as the number of KTs increases) applied to the multiple possible system configurations and COAs precludes detailed WBS budgeting for each. Furthermore, WBS budgeting is not done at the KT level. This makes estimation of KT-specific fixed and variable costs difficult, and simplifying assumptions may have to be used. We recommend that sensitivity analysis be used in cases in which model inputs are uncertain and that the user fully document the assumptions made in forming those inputs.
- RTRAM makes many assumptions about parameterization and functional form. We list the specific model assumptions in Appendix B.
- Performance is measured in a single dimension. RTRAM assumes that performance degradation at the system level is the sole measure of interest in terms of technical risk. This need not be the case. In some cases, multiple (objective) performance measures should be used in order to compare systems. This is likely case-specific. As documented in Chapters Three and Four, there are no theoretical constraints on adding multiple performance out-

comes; however, in practice, this extension would likely take considerable time and effort.

- A lack of correlation across input distributions (and parameters) is assumed. Garvey (2000) and Covert (2013), in discussing joint schedule and cost distributions, note that there are likely significant correlations between elements both between and within elements of each outcome. RTRAM's structure necessarily induces correlation between technical, schedule, and cost outcomes through assumed functional relationships, as detailed in Chapters Four and Five and in Appendix B. However, we assume that KT-specific schedule distributions are independent, which they may not be. Additionally, if parametric uncertainty were to be introduced, the correlation structure between parameter elements would be a key input into the model. Allowing for correlations would be a reasonably straightforward exercise once a proper method of simulating from correlated distributions is selected (see, e.g., Lurie and Goldberg, 1994).

- Aggregation to the system level involves very specific assumptions about the relationship between technologies and schedule. RTRAM takes technology-level inputs and stochastic outcomes and aggregates them to the alternative level to create distributions related to performance, schedule, and cost. As detailed in Chapters Four and Five, one must make assumptions about this process. For this version of the model, we assume that the performance consequence is the maximum of the individual KT consequences. In some cases, this assumption might not be warranted. So long as the user can make another defensible assumption about aggregating performance outcomes to the system level, relaxation of this structure is straightforward. Similarly, RTRAM aggregates schedules to the alternative level by using the rules documented in Chapter Five and Appendix B. Relaxation of these rules is straightforward as long as one can represent alternative rules mathematically.

- The model system boundaries are designed to encompass the acquisition process only through MS C. A more comprehensive trade space may include the entire acquisition process, including procurement and operations and maintenance. The extension into this space is conceptually straightforward.

Taken together, these limitations essentially imply that the outputs from any structural model, such as RTRAM, are only as good as the assumptions and data used to create them. Sensitivity analysis should be used to test robustness to various model assumptions in cases in which the analyst is uncertain about relevant inputs and their relationships.

Model Extensions

Most of the limitations listed above could be overcome with further research. We consider the following to be the most impactful for the Army in the short term.

Reconsider the Treatment of Technical Risk and Performance Outcomes in the Model

To make useful between-system comparisons, setting a normalized baseline across the performance dimension is essential. If subjective information from SMEs is to be used for this purpose, then changing the elicitation mechanism in the risk workshop will be necessary. The availability and use of objective performance information should also be explored.

Additional progress could also be made on the conceptualization of the counterfactual technologies implicitly assumed in the model. In some cases, there may be a functional relationship between performance metrics and time; in others, the performance outcome may be best viewed as a random variable.

Additionally, it seems likely that many weapon systems will be characterized by multiple performance dimensions and that summarizing this information in one dimension (say, by linear weighting of different performance metrics) may obscure information useful to decisionmakers. However, the expansion of dimensionality can also add complexity to the analysis and interpretation of results. Additional research is needed to explore the optimal structure of technical and performance information considered in the risk-adjusted trade-space model.

More Fully Integrate the Capabilities and Analysis of the Army's Cost Teams into the Structure of the Model

The source of stochastic variation in RTRAM is twofold: KTs and counterfactual technologies can be delivered or not delivered at the milestone date, and, depending on the COA and assumed technology-specific schedule distributions, this delivery time is a random variable as well. As noted elsewhere in this report, the WBS does not include schedule as a key component, and other representations of joint schedule/cost probability distributions, which inherently contain stochastic schedule outcomes, do not have the capability of estimating the impacts of behaviors intended to mitigate risk across technological, schedule, and cost dimensions.

There appears to be significant opportunity to incorporate additional information from the cost community into the risk-adjusted trade-space model and to develop best practices that could naturally lead to improved model outputs. Among these are more-detailed breakdowns of WBS-type categories across KTs (allowing for KT-specific marginal or average and fixed-cost parameters), the incorporation of stochastic cost parameters, and the introduction of potential correlation between these parameters.

Use Historical Data to Refine (and Perhaps Empirically Estimate) the Assumed Relationships in the Trade-Space Model

RTRAM is a structural model useful for investigating ex ante likelihoods of outcomes across performance, schedule, and cost dimensions. As such, it is based on functional relationships that we assume or that the user programs. The parameterization of an alternative is completely determined by the user.

To the extent that the data are available, opportunities exist to examine past programs and refine, and perhaps empirically estimate, some historical average relationships of the type assumed in the model. This could include information about realized counterfactual technologies, the relationships between fixed and marginal or average costs by KT for historical systems, and information about KT-specific schedule outcomes.

Continue Interaction with the Potential User Community to Identify Features That Would Be Useful to Add to the Model

The current RTRAM is a first step toward a hybrid structural/statistical model that conceptualizes the pre–MS C acquisition process as a set of stochastic outcomes across multiple, linked dimensions that allows users to quantitatively investigate potential actions that mitigate risk in one or more dimensions. We believe that this decision-support tool is a useful addition to current acquisition methodologies and provides a contribution in light of the Weapon Systems Acquisition Reform Act of 2009 and the needs of the user community.

However, the framework and tool are only as useful as the information provided by the model is useful to decisionmakers. Continued interaction with the community, and responsiveness to its demands for analysis, is essential to ensuring the best services to the warfighter as the fiscal environment continues to evolve.

User Manual for the Risk-Informed Trade Analysis Model Demonstration Tool, Version 1.0

The RTRAM demonstration tool (version 1.0) described in this document can be used in its existing application for visualizing a risk-informed trade space of the cost, schedule, and performance of a weapon system in an acquisition program. It acts as the UI for the underlying RTRAM and, in its current form, must be populated with data from AMSAA's risk workshop and technology-level variable and fixed costs.

Instructions for Using the Risk-Informed Trade Analysis Model Demonstration Tool

This section contains step-by-step instructions for a user of the RTRAM demonstration tool. The instructions lead the user through an exercise to explore the cost, schedule, and performance trade space surrounding a set of alternatives being evaluated by an acquisition program.

1. Open the Excel file, enabling macros.
2. On the **Alternatives** tab, enter the name of each alternative in a separate row. Note that an *alternative* is a collection of technologies and technology-level COAs. Therefore, one *system* may be configured into many different alternatives by switching the COAs of the system's KTs. Make sure to provide descriptive names to each alternative such that they may be easily distinguished in the output interface.

3. On the **Key Technologies** tab, enter the name of each technology on a separate row. In each of the subsequent columns, enter the values provided by the risk workshop. The first nine columns are the minimum, maximum, and most-likely values (in months) each, for time until the technology will reach TRL 7, MRL 8, and IRL 8, respectively. For the **Perf** column, enter the performance consequence from the risk workshop. Finally, enter the fixed and variable costs of each technology. Note that, because each row contains the fixed- and variable-cost parameters, the same physical subsystem with different cost characteristics should be defined as a different KT.

4. On the **Schedule Durations** tab, enter each of the schedule durations (in months) to be considered on a separate row.

Input User Interface

1. Select the **UI-1** tab, then choose **Re-Build UI** to populate the interface. See Figure A.1.

2. Once the interface has populated, design the alternatives to be used in RTRAM.

3. For each alternative, use the checkboxes to select the technologies to be considered.

4. Choose a COA for each technology that has been checked.

5. Use the **Select alternatives** and **Select durations** boxes to choose which alternatives to run for which milestones. RTRAM will run only the selected options.

6. After the alternatives have been designed, choose **Run RTRAM**. The MC analysis runs. Because of its use of memory, Excel may display a **Not Responding** message at the top of the program window; however, this is not a problem under normal circumstances.

Output User Interface: Risk Graph

1. Choose the **Histogram** tab. This UI provides numerous views of the schedule, cost, and performance (cumulative probability) distributions as a function of their consequences (i.e.,

Figure A.1
Input User Interface of the Risk-Informed Trade Analysis Model
Demonstration Tool

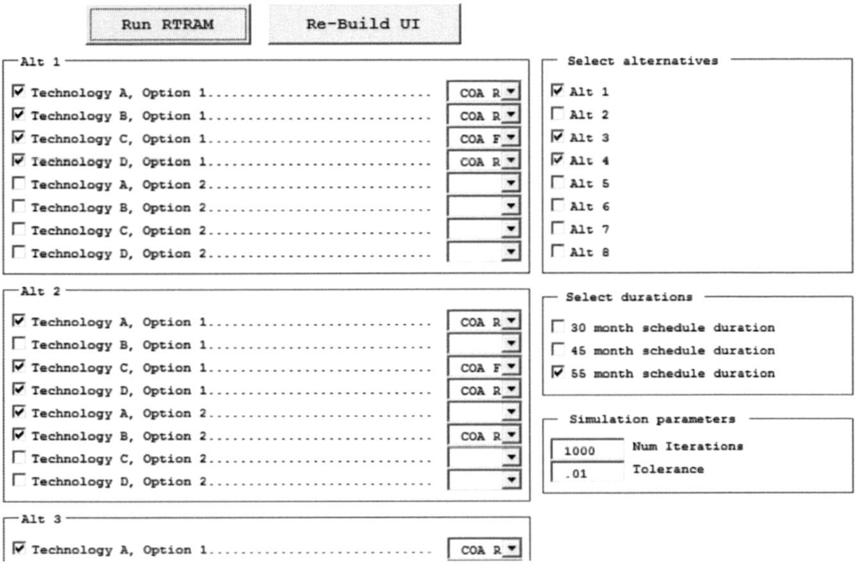

RAND *RR701-A.1*

months, dollars, and 0-to-5 degradation scale, respectively). See Figure A.2.

2. Use the **Select output** radio buttons at the top of the UI to choose the distribution (schedule, cost, or performance) to display on the graph.

3. In the upper right of the UI, choose the alternatives to display on the graph. You may choose one or multiple alternatives.

4. In the lower right, choose a schedule duration time.

5. If desired, below the graph, choose the constraints (performance, schedule, and cost) to be applied to the underlying data. If no constraints are desired, leave the drop-down boxes blank; otherwise, specify the constraint set of interest. Current functionality allows for one constraint per dimension.

6. At the top right, choose **Build Chart** to populate the graph.

Figure A.2
Risk-Graph User Interface of the Risk-Informed Trade Analysis Model Demonstration Tool

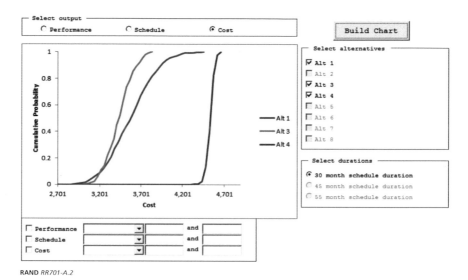

7. Change the dimension, schedule duration, alternatives chosen, and constraints as many times as desired, choosing **Build Chart** after each change.

Output User Interface: Three-Dimensional Trade Space

1. Choose the **Bubble Chart** tab. This interface provides a three-dimensional frontier of the mean schedule, mean cost, and mean performance values for each alternative under consideration. That is, each alternative's mean cost and performance (along with its confidence intervals) are plotted on a scatterplot, with the alternative's schedule displayed as a data-point label. See Figure A.3.
2. Choose a schedule duration and highlight all the alternatives chosen for display on the graph.
3. At the top right, choose **Build Chart** to populate the graph.
4. Change the schedule duration and alternatives chosen as many times as desired, choosing **Build Chart** after each change.

Figure A.3
Three-Dimensional Trade-Space User Interface of the Risk-Informed Trade
Analysis Model Demonstration Tool

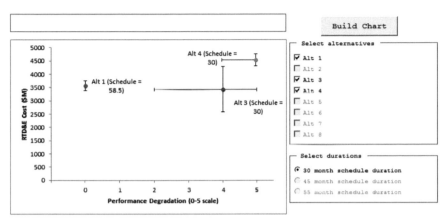

In addition to the preprogrammed output, RTRAM provides the user with the calculated performance, schedule, and cost draws corresponding to each MC iteration. These data are provided on the **Model Output** tab and include (across columns) (1) the alternative name, (2) the user-supplied schedule duration (e.g., the milestone date) for that run, (3) the MC iteration number, (4) the performance outcome (labeled **cPerf**), (5) the schedule outcome (labeled **cSched**), and (6) the cost outcome (labeled **cCost**). All outcome units are consistent with the units provided on the input tabs. The number of MC iterations can be changed by changing the assigned value of **intNumMCIterations** in RTRAM.

Outline of Specific Assumptions and Methods in the Risk-Informed Trade Analysis Model

This appendix describes the specific assumptions, including functional forms, parameterizations, and methodology, used in the Microsoft Excel–based RTRAM program delivered to AMSAA in September 2013. It provides a bridge between the mathematical framework presented in Chapter Five and the computer code detailed in Appendix D. The model was coded in VBA.

User-Provided Parameters

The **Alternatives** tab provides the user with the opportunity to name the systems of interest. This can be done before or after entering the information on the **Key Technologies** tab. These names will appear on the **UI-1** tab.

The **Key Technologies** tab is the primary source of user-provided data for RTRAM. For each possible KT under consideration, the user provides the following:

- the KT name
- the minimum value of time to TRL 7 for that technology, obtained through the risk workshop (TRL_{min})
- the maximum value of time to TRL 7 for that technology, obtained through the risk workshop (TRL_{max})
- the most likely value of time to TRL 7 for that technology, obtained through the risk workshop (TRL_{ml})

- the minimum value of time to MRL 8, conditional on achievement of TRL 7 for that technology, obtained through the risk workshop ($MRL_{\min}$)
- the maximum value of time to MRL 8, conditional on achievement of TRL 7 for that technology, obtained through the risk workshop ($MRL_{\max}$)
- the most likely value of time to MRL 8, conditional on achievement of TRL 7 for that technology, obtained through the risk workshop (MRL_{ml})
- the minimum value of time to IRL 8, conditional on achievement of TRL 7 for that technology, obtained through the risk workshop ($IRL_{\min}$)
- the maximum value of time to IRL 8, conditional on achievement of TRL 7 for that technology, obtained through the risk workshop ($IRL_{\max}$)
- the most likely value of time to IRL 8, conditional on achievement of TRL 7 for that technology, obtained through the risk workshop (IRL_{ml})
- the unidimensional measure of the performance consequence of KT nondelivery (or, equivalently, of counterfactual technology delivery), denoted $p_{nk}\left(kt_{nk}^{-}\right)$ in Chapter Five
- the total fixed cost associated with the KT, denoted fc_{nk}^{+} in Chapter Five
- the marginal or average cost per unit time for the KT, denoted vc_{nk}^{+} in Chapter Five[1]
- the total fixed cost associated with the counterfactual technology, denoted fc_{nk}^{-} in Chapter Five
- the marginal or average cost per unit time for the counterfactual technology, denoted vc_{nk}^{-} in Chapter Five.

The TRL, MRL, and IRL distributions, denoted $g_{nk}^{TRL}(t)$, $g_{nk}^{MRL}(t)$, and $g_{nk}^{IRL}(t)$ in Chapter Five, are assumed to be triangular. They refer to KT kt_{nk}^{+}, in that they describe the probability weights

---

[1]   Because marginal variable costs are assumed constant, marginal variable costs equal average variable costs.

associated with each potential readiness date of a programmatic KT. The distributions associated with the counterfactual technology kt_{nk}^- depend on the COAs (coa_{nk}) chosen across all technologies k for a given system n. In this version of RTRAM, the user cannot enter separate distributions for kt_{nk}^-.[2]

The user determines units for performance, schedule, and cost information, but the units should be consistent with the mathematical expressions detailed herein.

The **Schedule Duration** tab provides the user with the opportunity to enter the list of schedule durations, or assumed milestones, that can be chosen on the **UI-1** tab. These correspond to the milestone date *ms* in Chapter Five.

COAs, denoted coa_{nk} in Chapter Five, are chosen for each KT in each system on the **UI-1** tab via drop-down boxes.

See Chapter Six and Appendix A for more details about the UI.

Numerical Monte Carlo Analysis

RTRAM is built around the idea that, from a pre-milestone standpoint, the physical system itself (the collection of KTs and counterfactual technologies), the delivery date of that system, and the cost of that system are all stochastic. To numerically estimate the performance, schedule, and cost distributions associated with a planned system and set of behavioral assumptions about the acquisition process (the courses of action), RTRAM uses MC techniques to numerically simulate the outcomes. Briefly, the logic behind the model is as follows:

- For $N > 0$ iterations,
  - Take individual schedule draws for each KT.
  - Calculate the realized KT schedule outcomes.

---

[2] In other words, only the information on the programmatic KTs is utilized in the current version, in combination with the realized delivery time calculated from these primal distributions. In future versions, more-complicated specifications that allow for user-identified counterfactual distributions could be introduced.

- Aggregate the KT schedule outcomes to a system level to obtain a realized physical system and system-level delivery date (schedule outcome).
- Calculate the system-level performance outcome associated with the realized physical system.
- Calculate the system-level cost outcome associated with the realized physical system.

Compiling the performance, schedule, and cost outcomes into suitable ranges of each dimension and calculating frequency proportions within each range provide an estimate of the probability distribution of each outcome. These empirical distributions are the primary output from RTRAM and can be used in further analysis.

Unconditional Key Technology–Specific Schedule Distributions

The unconditional KT-specific schedule distributions $g_{nk}(t)$ are numerically estimated in RTRAM through pseudo-random sampling, assuming that the TRL, MRL, and IRL distributions from the risk workshop are (conditionally) independent. These distributions are assumed to be triangular, such that

$$
g_{nk}^{TRL}(t) = \begin{cases}
0 \text{ for } t < TRL_{min}, \\
\dfrac{2(t - TRL_{min})}{(TRL_{max} - TRL_{min})(TRL_{ml} - TRL_{min})} \text{ for } TRL_{min} \le t \le TRL_{ml}, \\
\dfrac{2(TRL_{max} - t)}{(TRL_{max} - TRL_{min})(TRL_{min} - TRL_{ml})} \text{ for } TRL_{ml} < t \le TRL_{max}, \\
\text{and} \\
0 \text{ for } t > TRL_{max}
\end{cases}
$$

$$
g_{nk}^{MRL}(t) = \begin{cases}
0 \text{ for } t < MRL_{min}, \\
\dfrac{2(t - MRL_{min})}{(MRL_{max} - MRL_{min})(MRL_{ml} - MRL_{min})} \\
\text{for } MRL_{min} \leq t \leq MRL_{ml}, \\
\dfrac{2(MRL_{max} - t)}{(MRL_{max} - MRL_{min})(MRL_{min} - MRL_{ml})} ; \text{ and} \\
\text{for } MRL_{ml} < t \leq MRL_{max}, \\
\text{and} \\
0 \text{ for } t > MRL_{max}
\end{cases}
$$

$$
g_{nk}^{IRL}(t) = \begin{cases}
0 \text{ for } t < IRL_{min}, \\
\dfrac{2(t - IRL_{min})}{(IRL_{max} - IRL_{min})(IRL_{ml} - IRL_{min})} \\
\text{for } IRL_{min} \leq t \leq IRL_{ml}, \\
\dfrac{2(IRI_{max} - t)}{(IRL_{max} - IRL_{min})(IRL_{min} - IRL_{ml})} \\
\text{for } IRL_{ml} < t \leq IRL_{max}, \text{ and} \\
0 \text{ for } t > IRL_{max}
\end{cases}
$$

Let t_{TRL} be a pseudo-random draw from the triangular TRL distribution and t_{MRL} and t_{IRL} be corresponding pseudo-random draws from the implied MRL and IRL distributions, respectively. We assume that these draws (and the distributions) are independent. The technology readiness date t_{nk} is assumed to be $t_{nk} = t_{TRL} + \max(t_{MRL}, t_{IRL})$, which is a single draw from $g_{nk}(t)$. This draw represents one realized time at which the KT achieves TRL 7, MRL 8, and IRL 8. The compiled draws, corresponding to the empirical KT-specific schedule distributions, are reported on the **Distributions** tab.

Nine States of Key Technologies Are Represented in the Model

Table B.1 illustrates the nine states of the KTs in RTRAM. Throughout the discussion in this section, we reference the specific state number, shown in the last column in the table.

Key Technology–Specific Schedule Outcomes

The KT-specific schedule draws are converted to KT-specific outcomes rt_{nk} using the milestone date ms, the COAs, coa_{nk}, and the schedule draws t_{nk} themselves. If $coa_{nk} \neq coa^e \forall k$, then the readiness date of the kth technology for the nth alternative is determined by $rt_{nk} = \min(t_{nk}, ms)$ (states 6 through 9). If $coa_{nk} = coa^e$ for at least one k, then $rt_{nk} = \min\left(\max\left(\mathbf{t}_{n(-e)}\right), t_{nk}\right)$ (states 1 through 5), where $\max(\mathbf{t}_{n(-e)})$ is the maximum schedule draw from all technologies with coa^e. This map from draws to outcomes jointly determines the physical system that is delivered in terms of the portfolio of KTs and counterfactual technologies and the system delivery date, as described in the next section.

Aggregation of Key-Technology Schedule Outcomes to the Alternative Level

Physical System

Each set of readiness times $\{rt_{n1}, \ldots, rt_{nK}\}$ for one draw of alternative n maps directly to inclusion or noninclusion of a KT or counterfactual technology in the final system, based on the COAs coa_{nk} and the readiness time rt_{nk}. This map corresponds to the function $a\left(\mathbf{kt}_n^i\right)$ in Chapter Five. As such, the calculated readiness times are used to calculate one draw from $g_n^a\left(\mathbf{kt}_n^i\right)$, described as follows.

Assume that, for all k, $coa_{nk} \neq coa^e$. If $rt_{nk} = t_{nk}$, then the KT is assumed to be ready before or on the milestone date and is delivered at the system delivery time (states 6 and 8). If $rt_{nk} = ms \neq t_{nk}$ and

Table B.1
Courses of Action, Draw Conditions, and Technology-Specific Schedule, Performance, and Cost Outcomes of the Risk-Informed Trade Analysis Model

Behavioral Assumption	Alternative COA coa_{nk}	Draw Condition		Technology-Specific Outcome			State
		Delivered Technology	Schedule Condition	Schedule	Performance	Cost	
At least $coa^e \forall k$. In this case, $$rt_n = \max \left\{ \begin{array}{l} \max(\mathbf{t}_{n(\cdot)}), \\ \min\left(\begin{array}{l} \max(\mathbf{t}_{n(-k)}), \\ ms \end{array} \right) \end{array} \right.$$	coa^e	KT	$t_{nk} \le rt_n$	t_{nk}	0	$c_{nk}^+ = fc_{nk}^+ + vc_{nk}^+ \times t_{nk}$	1
	coa^f	KT	$t_{nk} \le rt_n$	t_{nk}	0	$c_{nk}^+ = fc_{nk}^+ + vc_{nk}^+ \times t_{nk}$	2
		KT	$t_{nk} > rt_n$	rt_n	0	$c_{nk}^+ = fc_{nk}^+ + vc_{nk}^+ \times t_{nk}$	3
	coa^r	KT	$t_{nk} \le rt_n$	t_{nk}	0	$c_{nk}^+ = fc_{nk}^+ + vc_{nk}^+ \times t_{nk}$	4
		Counterfactual	$t_{nk} > rt_n$	r_{tn}	p_{nk}	$c_{nk}^+ = fc_{nk}^+ + vc_{nk}^+ \times t_{nk}$	5

Table B.1—Continued

Behavioral Assumption	Draw Condition			Technology-Specific Outcome			
Alternative COA	coa_{nk}	Delivered Technology	Schedule Condition	Schedule	Performance	Cost	State
No $coa^e \forall k$. In this case, $rt_n = \min\{\max(\mathbf{t}_{nk}), ms\}$	coa^f	KT	$t_{nk} \leq rt_n$	t_{nk}	0	$c_{nk}^+ = fc_{nk}^+ + vc_{nk}^+ \times t_{nk}^+$	6
		KT	$t_{nk} > rt_n$	rt_n	0	$c_{nk}^+ = fc_{nk}^+ + vc_{nk}^+ \times t_{nk}^+$	7
	coa^r	KT	$t_{nk} \leq rt_n$	t_{nk}	0	$c_{nk}^+ = fc_{nk}^+ + vc_{nk}^+ \times t_{nk}^+$	8
		Counterfactual	$t_{nk} > rt_n$	rt_n	p_{nk}	$c_{nk}^+ = fc_{nk}^+ + vc_{nk}^+ \times rt_{nk}^+ + fc_{nk}^-$	9

NOTE: See text for COA (coa) interpretation. rt_n = technology readiness time for alternative n. t_{nk}^+ = draw from combined TRL, IRL, and MRL distribution of technology k. $\mathbf{t}_{n(s)}$ = set of technologies associated with coa^s. $\mathbf{t}_{n(-s)}$ = set of technologies with COAs other than coa^s. ms = milestone date. c_{nk} = total technology cost, with + representing KT and – representing counterfactual technology. fc = fixed cost. vc = variable cost. Only feasible outcomes are shown.

$coa_{nk} = coa^r$, then the KT is assumed not ready at the milestone date and the counterfactual technology is delivered at the milestone date (state 9). If $rt_{nk} = ms \neq t_{nk}$ and $coa_{nk} = coa^f$, then the KT is assumed ready (and delivered with the entire system) at the milestone state (but will incur additional costs) (state 7).

Now assume that, for at least one k, $coa_{nk} = coa^s$. If $rt_{nk} = \max(\mathbf{t}_{\mathbf{n(e)}}) \neq t_{nk}$ and $coa_{nk} = coa^r$, then the KT is assumed to not be ready at $\max(\mathbf{t}_{\mathbf{n(e)}})$ and the counterfactual technology is assumed delivered at this date (state 5). If $rt_{nk} = \max(\mathbf{t}_{\mathbf{n(e)}})$ and $coa_{nk} = coa^e$ or $coa^{nk} = coa^f$, then the kth technology's draw is the maximum of all the draws and the KT will be ready (and delivered with the entire system) at this date (state 1 or 3). If $rt_{nk} = \max(\mathbf{t}_{\mathbf{n(e)}}) \neq t_{nk}$ and $coa_{nk} = coa^f$, then the KT is assumed ready (and delivered with the entire system) at $\max(\mathbf{t}_{\mathbf{n(e)}})$ with additional costs (state 3). If, under this assumption, $rt_{nk} \geq t_{nk}$, then the KT is assumed ready at t_{nk} (state 3). If the draw happens to be no greater than rt_{nk} regardless of COA, then the technology is assumed ready at t_{nk} (states 1, 2, and 4).

Alternative Delivery Date or Schedule Outcome

For any draw of the MC procedure, the physical system is assumed delivered in that the planning horizon is over, at a particular date r_n, which is assumed to be the schedule outcome for that draw. Chapter Five documents the calculation of this date in terms of t_{nk}. An equivalent specification is $rt_n = \max\{rt_{n1}, \ldots, rt_{nK}\}$. This represents one draw from the distribution $g_n(t_{n1}, \ldots, t_{nK})$ defined in Chapter Five and assumes that the schedule distributions of the KTs are independent.

Alternative Performance Outcome, by Course of Action

Once a physical system has been constructed from the probability draws, it has a deterministic relationship with the performance outcome through the technology-specific performance outcomes $p_{nk}(kt_{nk}^{-})$ entered by the user (realized in states 5 and 9). We assume that this map, denoted P_n in Chapter Five, is $P_n = p_n\left(p_{n1}(kt_{n1}), \ldots, p_{nK}(kt_{nK})\right) = \max\left(p_{n1}(kt_{n1}), \ldots, p_{nK}(kt_{nK})\right)$. As such, users should be aware that increasing values of $p_{nk}(kt_{nk}^{-})$ correspond to increased performance degradation. Each P_n is a draw from the implied system performance distribution.

This version of RTRAM has been developed such that delivery of a KT is coded as $p_{nk}\left(kt_{nk}^+\right) = 0$ (states 1 through 4 and 6 through 8). As such, differentials in performance across delivered KTs are not explicitly considered. Furthermore, the standard output graphs assume that performance degradation spans the integer range 1 through 5, an assumption that is consistent with results of the risk workshop. Relaxing this assumption can be accomplished by changing the relevant portions of the code.[3]

Alternative Cost Outcome, by Course of Action

A physical system coupled with each KT or counterfactual readiness time $\{rt_{n1}, \ldots, rt_{nK}\}$ provides the information necessary to estimate costs for the physical system. As described in Chapter Five, we assume that total alternative costs are the sum of total fixed costs and total variable costs. Total fixed costs are those system costs not expected to vary across the time horizon modeled by RTRAM but that may vary across the delivered technology (i.e., the KT or the counterfactual). Total variable costs consist of all costs expected to be incurred per unit time. Total alternative-level fixed and variable costs are calculated from their technology-specific counterparts.

The exact form of the coded cost functions depends on the delivered technology. If a KT is delivered, then total costs associated with the kth technology are $c_{nk}^+ = fc_{nk}^+ + vc_{nk}^+ \times rt_{nk}$ (states 1 through 4 and 6 through 8). If the counterfactual technology is delivered, then total costs associated with the kth technology are $c_{nk}^- = fc_{nk}^+ + vc_{nk}^+ \times rt_{nk} + fc_{nk}^-$ (states 5 and 9). Total alternative-level costs are the sum of the technology-specific fixed and variable costs.

The information necessary to obtain the fixed- and marginal variable–cost components is typically not readily available from standard analyses. As such, the user must either estimate or assume these values. For example, in Chapter Six, the analysis assumes 100-percent variable costs and uses only information on projected RDT&E costs

---

[3] We strongly recommend reconsidering these assumptions.

and SME schedule distributions to apportion costs by technology. In other cases, additional information may be available. For example, a detailed analysis of the WBS cost elements of a planned system could include apportioning each element to either fixed (not varying with time) or variable (varying with time) categories and making some assumptions about the expected time to completion (in the case of variable costs). There are many other options, depending on the availability of cost information. For the purposes of incorporation into RTRAM, however, the parameterization requires only assumptions about fixed and marginal variable costs.

Expert Elicitation and Minimizing Bias from Heuristics

Expert elicitation is a systematic process for obtaining information from experts about specific uncertain quantities. The process includes explicit criteria for expert selection and a detailed interview or focus-group protocol (i.e., an instruction manual for the facilitator with recommended language or prompts). Additionally, the expert elicitation may include briefing materials for (or a workshop with) experts that are presented before the actual elicitation, providing details about the subject matter and the elicitation process itself (Morgan and Henrion, 1990).

During the elicitation, experts are essentially asked to make subjective judgments about the uncertain quantities of interest. Although each expert will likely use his or her best judgment that has been informed by his or her historical experience with the subject matter, each will nevertheless make use of heuristic processes. Heuristics are mental shortcuts people use to make quick judgments to ease the cognitive load of making decisions (Hastie and Dawes, 2010; Ayyub, 2001b; Galway, 2007; Kahneman, Slovic, and Tversky, 1982; Morgan and Henrion, 1990). Because time and resources are finite, people develop simple processes that allow them to make decisions of varying complexity in a relatively short amount of time using only their current knowledge.

Every person uses heuristic processes to make decisions on a daily basis. Indeed, as expressed by Tvserky and Kahneman (1974, p. 1124), they "are quite useful, but sometimes . . . lead to severe and systematic errors." It is those errors that are of concern when conducting an expert elicitation. However, having knowledge about and strategies for mitigating those heuristics that are most likely to bias an expert elicitation can greatly improve the elicitation and the reliability of its results.

Table C.1 lists six heuristic processes that may commonly be used by experts who are estimating acquisition- and risk-related quantities of schedule, cost, and performance outcomes. The table provides a description and an example for each heuristic.

Table C.1
Heuristics, Adapted from Hastie and Dawes, 2010

Heuristic	Description	Example
Availability	Tendency to overestimate probability of events that are easy to recall	Frequency of deaths from shark attacks versus deaths from falling airplane parts. The former is always in the media and is therefore more available to our memory. However, the latter occurs more frequently.
Representativeness	Judging probability of events by focusing on (potentially irrelevant) characteristics of other events that are similar; neglecting information about base rates (i.e., the unconditional frequency of an event occurring)	Description of a student from Massachusetts Institute of Technology as female, traveled extensively, fluent in many languages, writes sonnets, and so on. Would you guess that she is an art-history major or an engineer? Base rates would lead to the conclusion of engineer, but the representative heuristic could cause someone to believe that art history is more likely.
Anchoring and adjusting	Biasing of a final assessment value toward an initial anchor value by constraining adjustment in light of new evidence	If first asked, "Is the population of the Washington, D.C., metro area more or less than 1 million?" then, if asked, "What is the population of the D.C. metro area?" one may anchor and insufficiently adjust based on the anchor of 1 million provided.
Overconfidence	Underestimation of uncertainty about a quantity	Most people will believe that they are better-than-average drivers, but, in any sufficiently large sample, about half of the drivers are worse than average.

Table C.1—Continued

Heuristic	Description	Example
Conjunction fallacy	Assuming that specific conditions are more probable than a single general one	A woman is 35 years old, outspoken, single, and very bright. As a student, she majored in philosophy, was deeply concerned about issues of discrimination and social justice, and participated in antinuclear demonstrations. Which is more probable? That the woman is a bank teller or that she is a bank teller *and* is active in the feminist movement? The single general case (bank teller only) is always more probable than the specific joint case.
Hindsight bias	Being inclined to see events that have already occurred as being more predictable than they were before they took place	After viewing the outcome of a potentially unforeseeable event, a person believes that he or she "knew it all along."

Mitigating Bias from Heuristics

In part to mitigate potential bias from heuristics, researchers and practitioners in the field have developed a variety of best practices for expert elicitation. The best practices can be very specific to what is being elicited, who is being asked, and how the elicitation results will be used. However, most (if not all) good elicitations have a few things in common.

First, and most importantly, a proper elicitation follows a systematic elicitation method. This will benefit the elicitation by making it more reproducible, more easily documentable, and more controllable (Morgan and Henrion, 1990; Ayyub, 2001a). First, a systematic method will be easier to reproduce. Although elicited quantities should not take the place of real data, they are still desirable to be able to reproduce results. A reproducible process should lead to some stability in elicited opinions between different elicitation facilitators or over time (but not between different experts). A systematic process will also be easier to document. That is, if a detailed protocol exists, it will guide documentation so that elicitation results are recorded in an easily comprehensible and auditable format. Finally, systematic elicitation methods allow the facilitator to have more control over the topic being considered, as well as over the experts themselves. An elicitation protocol can be designed to a desired level of control over expert answers (Morgan and Henrion, 1990). These can range from completely structured (e.g., only elicit a specific value) to semistructured (e.g., elicit a value, as well as rationale for that value) to open-ended methods (e.g., use open-ended questions, directing the discussion to areas of interest through probes only if necessary). A protocol will benefit from being pilot-tested with an expert (or doing a dry run of the protocol with a set of experts) to ensure that it is understandable and reasonable. As a part of this pilot test, or during the expert elicitation itself, unambiguous definitions of the quantities to be elicited should be developed (Ayyub, 2001a; Morgan and Henrion, 1990).

Many researchers (e.g., Ayyub, 2001b; Morgan and Henrion, 1990) suggest that, prior to the actual elicitation, experts be trained. This training may involve providing detailed background materials about the subject matter in question, teaching experts about known heuristics that may bias their elicitation answers, and a calibration exercise with experts that involves, for instance, eliciting a set of known probabilities and then allowing experts to compare their subjective assessments with the actual probabilities.

Furthermore, allowing experts to discuss their subjective judgments among themselves and then to revise these judgments based on the discussion is commonly cited as an important best practice (Ayyub, 2001a; Morgan and Henrion, 1990; Sackman, 1974; Armstrong, 2001). One benefit of this approach is that the discussion could broaden the range of opinions under consideration by an expert, which may, in turn, increase the range of values they would consider plausible. The approach may be especially helpful if multiple, independent, and heterogeneous experts are used and if experts are asked to provide, at a minimum, upper, lower, and most-likely values for the quantity under consideration (Galway, 2007). However, the protocol should never begin by eliciting the most likely value because this could anchor experts toward a central value and not prompt them to consider the tails of a distribution. Studies have also show that eliciting values as frequencies (e.g., one in 100) or odds (e.g., 1:100) rather than probabilities (e.g., 0.01) may improve experts' assessments of likelihoods (Galway, 2007; Morgan and Henrion, 1990).

Finally, detailed documentation will allow for clear communication of results, as well as any limitations of the elicitation. The documentation will also allow the elicited quantities to be archived, which may be revisited for future retrospective studies (Morgan and Henrion, 1990).

General Elicitation Protocols

Based on the best practices set out previously, a few general elicitation protocols are available that may be tailored to the specific needs of the elicitation. One protocol for eliciting subjective probability distributions may include the following elements (Morgan and Henrion, 1990; Ayyub, 2001a):

1. Elicit extreme values first.
2. Ask for *scenarios* that could lead to outcomes outside of the extreme values (counteract overconfidence). Iterate on extreme values if necessary.
3. Elicit probabilities for multiple values within the range (e.g., 25th, 50th, and 75th percentiles); choose values in random order to counteract anchoring bias.
4. Plot points on a cumulative distribution; do not show the plot to experts until all values have been elicited.
5. Verify the curve with experts; iterate to smooth the curve if necessary.

Protocols for groups of experts provide a means of structuring the group communication process. For example, experts can communicate through open discussions, structured discussions (sometimes referred to as the nominal group technique), or anonymous surveys (i.e., the Delphi method) (e.g., Ayyub, 2001b; Brown, 1968; Sackman, 1974).

Most group protocols, regardless of how experts communicate, follow the same four steps:

1.   Elicit values and rationale from individual experts.
2.   Aggregate results (e.g., mininum, maximum, mean) to provide (anonymous) feedback.
3.   Have experts revise values individually based on feedback and explain their rationale for extreme values.
4.   Iterate steps 2 and 3 until responses begin to converge (generally three rounds).

Expert Disagreement

If using a heterogeneous set of experts, they will likely disagree on the quantities that have been elicited, and converging on one value may seem impossible. When this happens, it is important to take a few very simple steps. First, understand why the experts disagree. Disagreement may be caused by their interpretations of the elicitation question or quantity, illuminating a need to reformulate the question or the definition of the elicited value. Second, allow for experts to interact, seeing whether differences can be reduced (e.g., using the Delphi method). Resolution of differences in this manner can often lead to better overall results (Morgan and Henrion, 1990). If consensus cannot be reached, the facilitator then has two options. The more preferred option is to treat the assessments that use the elicited values as inputs parametrically. This will help to understand output results from a range of expert opinions. Otherwise, a facilitator may decide to combine opinions into averages or weighted averages or use a Bayesian approach (Morgan and Henrion, 1990).

A large literature exists on different mathematical and administrative means of combining opinions (Armstrong, 2001; Ayyub, 2001a, 2001b; Clemen, 1989; Clemen and Winkler, 1999). Here, we mention a few rather simple ones to provide the conceptual basis for this task. A basic mathematical consideration is the complexity of the aggregation function. Much of the literature suggests that linear models (i.e., averages of some kind) work better than other models or intuition (Armstrong, 2001; Hastie and Dawes, 2010; Dana and Dawes, 2004). However, study results disagree about whether a weighted average is better than unweighted. For example, some show the opposite to be true (e.g., Hammitt and Zhang, 2013). If a weighted average is chosen, there are administrative considerations in choosing the weights. Weights may be chosen through experts' self-ratings of their expertise on the subject matter or a facilitator's rating of experts' expertise, or seed questions may be used to test the calibration or quality of each expert's judgment (Ayyub, 2001b). For the latter, if an expert's judgment calibrates well to the actual quantity, that expert's opinion may be weighted more heavily.

Applying the Literature to U.S. Army Materiel Systems Analysis Activity Risk Workshop Procedures

The literature presented thus far may be applied in different ways to continue to improve AMSAA's risk workshop process and mitigate biases stemming from experts' heuristics. First, the systematic nature of the workshop may be improved by providing a very detailed protocol for the workshop facilitator, including the full script he or she should use. This may then be complemented by a documentation effort for each workshop in which the protocol is used as the organization structure for the documentation. Second, briefing materials could be provided to all SMEs in advance of the workshop about the subject matter (e.g., technologies), the elicitation process they will undergo, and a review of heuristics that could bias their opinions.

To address the heuristic of overconfidence that may be prevalent during the elicitation of transition times (e.g., time until TRL 7) and consequences, SMEs could be asked to think of scenarios in which the likelihood or consequence falls outside of the range of values first elicited. If SMEs can imagine such scenarios, the original range considered would likely need to be expanded. Additional elicitation processes can be included to provide a better range of transition times. These include first eliciting the maximum and minimum transition times, then randomly eliciting 25th-, 50th-, and 75th-percentile values. This cumulative distribution curve can then be plotted, and experts may iteratively refine and smooth the distribution. Furthermore, instead of eliciting conditional values for transition times (e.g., time until technology reaches IRL 8, after it has reached TRL 7), elicit those for only independent events. Conditional values are cognitively more difficult than unconditional ones.

Finally, procedures can be implemented to address the group dynamics that occur during the risk workshop. For instance, a facilitator may first conduct individual surveys to capture experts' initial assessments. This could be completed in a survey before the workshop has begun. Statistics (e.g., means and standard deviations) could be calculated based on these surveys and presented to experts when they first arrive for the workshop. A facilitator can then choose from some of the procedures previously mentioned to attempt to reach group consensus. If no consensus can be reached, averages may be calculated. If a weighted average is chosen, a set of seeding or calibration questions may be used to assign weights to experts.

Computer Code

The RTRAM code is presented in this appendix. In Microsoft Excel, this code can be accessed by clicking the **Visual Basic** button of the **Developer** tab. For information about using VBA in Excel, search for "visual basic" in Microsoft Office's help system.

The RTRAM code is split into five major modules. The RTRAM module contains the MC simulation code.

In VBA, a comment is indicated by a single quotation mark.

BuildUI Module

This module takes the user input and builds the UI on the **UI-1** tab.

```
BuildUI - 1

Option Explicit


Sub RemoveAllControls(ws As Worksheet)
    Dim oleObj As Object
    For Each oleObj In ws.OLEObjects
        If oleObj.progID = "Forms.Image.1" Then
            oleObj.Delete
        End If

        If oleObj.progID = "Forms.Label.1" Then
            oleObj.Delete
        End If

        If oleObj.progID = "Forms.CheckBox.1" Then
            oleObj.Delete
        End If

        If oleObj.progID = "Forms.ComboBox.1" Then
            oleObj.Delete
        End If

        If oleObj.progID = "Forms.OptionButton.1" Then
            oleObj.Delete
        End If

        If oleObj.progID = "Forms.TextBox.1" Then
            oleObj.Delete
        End If
    Next oleObj
End Sub


Sub BuildInputUI()
    Dim i As Integer, j As Integer
    Dim row As Integer
    Call setSheets

    ' stop the flickering
    Application.ScreenUpdating = False

    ' clear out UI-1
    Call RemoveAllControls(wsInputUI1)

    ' get count of alternatives
    Dim NumAlternatives As Integer
    With wsAlternatives
        NumAlternatives = .Cells(Rows.Count, "A").End(xlUp).row - 1
    End With
    'MsgBox "Number of alternatives: " & NumAlternatives

    ' get count of key technologies
    Dim NumTechnologies As Integer
    With wsTechnologies
        NumTechnologies = .Cells(Rows.Count, "A").End(xlUp).row - 1
    End With
    'MsgBox "Number of key technologies: " & NumTechnologies

    ' get count of durations
    Dim NumDurations As Integer
    With wsDurations
        NumDurations = .Cells(Rows.Count, "A").End(xlUp).row - 1
    End With
    'MsgBox "Number of schedule durations: " & NumDurations

    ' ERROR checking before we begin
    Dim Duration As String
    For i = 1 To NumDurations
        Duration = wsDurations.Cells(i + 1, "A").Value
        If Not IsNumeric(Duration) Then
            MsgBox "Durations must be numeric: " & Duration
```

```
BuildUI - 2
            End
        End If
    Next i

    For i = 1 To NumTechnologies
        Dim Technology As String
        Technology = wsTechnologies.Cells(i + 1, "A").Value

        For j = 2 To 15
            If Not IsNumeric(wsTechnologies.Cells(i + 1, j).Value) Then
                MsgBox wsTechnologies.Cells(1, j).Value & " must be numeric: " & wsTechnologies.Cells(
i + 1, j).Value
            End
            End If
        Next j

        If wsTechnologies.Cells(i + 1, "B").Value > wsTechnologies.Cells(i + 1, "D").Value Or _
            wsTechnologies.Cells(i + 1, "D").Value > wsTechnologies.Cells(i + 1, "C").Value Then
            MsgBox Technology & ": check TRL parameters to ensure that Min <= ML <= Max"
            End
        End If

        If wsTechnologies.Cells(i + 1, "E").Value > wsTechnologies.Cells(i + 1, "G").Value Or _
            wsTechnologies.Cells(i + 1, "G").Value > wsTechnologies.Cells(i + 1, "F").Value Then
            MsgBox Technology & ": check MRL parameters to ensure that Min <= ML <= Max"
            End
        End If

        If wsTechnologies.Cells(i + 1, "H").Value > wsTechnologies.Cells(i + 1, "J").Value Or _
            wsTechnologies.Cells(i + 1, "J").Value > wsTechnologies.Cells(i + 1, "I").Value Then
            MsgBox Technology & ": check IRL parameters to ensure that Min <= ML <= Max"
            End
        End If

    Next i

    ' This builds this alternatiaves check and dropdown boxes
    '------------------------------------------------------------------------
    Dim Left As Double, Right As Double, Top As Double, Bottom As Double

    Dim GroupBoxHeight As Double, GroupBoxWidth As Double, CheckBoxWidth As Double, ComboBoxWidth As D
ouble
        GroupBoxHeight = 15 * NumTechnologies + 20
        GroupBoxWidth = 345
        CheckBoxWidth = 280
        ComboBoxWidth = 50

    row = 2
    For i = 1 To NumAlternatives
        Dim Alternative As String
        Alternative = wsAlternatives.Cells(i + 1, "A").Value

        Left = 50
        Top = 50 + (GroupBoxHeight + 10) * (i - 1)

        ' Create group box for alternative
        Dim gb As Object
        Set gb = wsInputUI1.OLEObjects.Add(ClassType:="Forms.Image.1", Left:=Left, Top:=Top, Width:=Gr
oupBoxWidth, Height:=GroupBoxHeight)
        With gb.Object
            .BackColor = RGB(256, 256, 256)
        End With

        Dim LabelWidth As Double
        LabelWidth = Len(Alternative) * 6
        Set gb = wsInputUI1.OLEObjects.Add(ClassType:="Forms.Label.1", Left:=Left + 10, Top:=Top - 5,
Width:=LabelWidth, Height:=15)
        With gb.Object
            .Font.Name = "Courier New"
            .Font.Size = 8
            .TextAlign = fmTextAlignCenter
```

```
BuildUI - 3
            .Caption = Alternative
        End With

        For j = 1 To NumTechnologies
            Technology = wsTechnologies.Cells(j + 1, "A").Value

            ' Create check boxes for technologies
            Dim cb As Object
            Set cb = wsInputUI1.OLEObjects.Add(ClassType:="Forms.CheckBox.1", Left:=Left + 5, Top:=(To
p + 13) + 15 * (j - 1), Width:=CheckBoxWidth, Height:=15)
            With cb
                .Name = "AT: " & i & ", " & j
            End With
            With cb.Object
                .Font.Name = "Courier New"
                .Font.Size = 8
                .Caption = PadRight(Technology, 50, ".")
                .Value = False
            End With

            ' Create combo boxes for COAs
            Set cb = wsInputUI1.OLEObjects.Add(ClassType:="Forms.ComboBox.1", Left:=Left + CheckBoxWid
th + 10, Top:=(Top + 13) + 15 * (j - 1), Width:=ComboBoxWidth, Height:=15)
            With cb
                .Name = "AT COA: " & i & ", " & j
            End With
            With cb.Object
                .Font.Name = "Courier New"
                .Font.Size = 8
                .AddItem "COA E"
                .AddItem "COA F"
                .AddItem "COA R"
            End With

            row = row + 1
        Next j
    Next i

    Right = Left + GroupBoxWidth

    ' This builds this alternatiaves to run box
    '------------------------------------------------------------------------
    Dim AltBoxLabel As Variant

    ' Create group box for alternatives
    Left = Right + 15
    Top = 50
    GroupBoxHeight = 15 * NumAlternatives + 20
    GroupBoxWidth = 205
    CheckBoxWidth = GroupBoxWidth - 10

    Set gb = wsInputUI1.OLEObjects.Add(ClassType:="Forms.Image.1", Left:=Left, Top:=Top, Width:=GroupB
oxWidth, Height:=GroupBoxHeight)
    With gb.Object
        .BackColor = RGB(256, 256, 256)                              .
    End With

    AltBoxLabel = "Select alternatives"
    LabelWidth = Len(AltBoxLabel) * 6
    Set gb = wsInputUI1.OLEObjects.Add(ClassType:="Forms.Label.1", Left:=Left + 10, Top:=Top - 5, Widt
h:=LabelWidth, Height:=15)
    With gb.Object
        .Font.Name = "Courier New"
        .Font.Size = 8
        .TextAlign = fmTextAlignCenter
        .Caption = AltBoxLabel
    End With

    ' Create check boxes for alternatives
    row = 2
    For i = 1 To NumAlternatives
```

```
BuildUI - 4

    Alternative = wsAlternatives.Cells(i + 1, "A").Value

    Set cb = wsInputUI1.OLEObjects.Add(ClassType:="Forms.CheckBox.1", Left:=Left + 5, Top:=(Top +
13) + 15 * (i - 1), Width:=CheckBoxWidth, Height:=15)
    With cb
        .Name = "A: " & i
    End With
    With cb.Object
        .Font.Name = "Courier New"
        .Font.Size = 8
        .Caption = PadRight(Alternative, 33, " ")
        .Value = False
    End With
    row = row + 1
Next i

Bottom = Top + GroupBoxHeight

' This builds this durations to run box
'------------------------------------------------------------------------

Left = Right + 15
Top = Bottom + 15
GroupBoxHeight = 15 * NumDurations + 20
GroupBoxWidth = 205
CheckBoxWidth = 195

' Create group box for durations
Set gb = wsInputUI1.OLEObjects.Add(ClassType:="Forms.Image.1", Left:=Left, Top:=Top, Width:=GroupB
oxWidth, Height:=GroupBoxHeight)
With gb.Object
    .BackColor = RGB(256, 256, 256)
End With

AltBoxLabel = "Select durations"
LabelWidth = Len(AltBoxLabel) * 6
Set gb = wsInputUI1.OLEObjects.Add(ClassType:="Forms.Label.1", Left:=Left + 10, Top:=Top - 5, Widt
h:=LabelWidth, Height:=15)
With gb.Object
    .Font.Name = "Courier New"
    .Font.Size = 8
    .TextAlign = fmTextAlignCenter
    .Caption = AltBoxLabel
End With

row = 2
For i = 1 To NumDurations
    Duration = wsDurations.Cells(i + 1, "A").Value

    ' ERROR checking
    If Not IsNumeric(Duration) Then
        MsgBox "Durations must be numeric: " & Duration
        End
    End If

    ' Create check boxes for durations
    Set cb = wsInputUI1.OLEObjects.Add(ClassType:="Forms.CheckBox.1", Left:=Left + 5, Top:=(Top +
13) + 15 * (i - 1), Width:=CheckBoxWidth, Height:=15)
    With cb
        .Name = "D: " & i
    End With
    With cb.Object
        .Font.Name = "Courier New"
        .Font.Size = 8
        .Caption = Duration & " month schedule duration"
        .Value = False
    End With
    row = row + 1
Next i

Bottom = Top + GroupBoxHeight
```

```
BuildUI - 5


    ' Create input for number of Monte Carlo iterations and tolerance
    '------------------------------------------------------------------------
    Dim TextBoxWidth As Double
    TextBoxWidth = 50

    Left = Right + 15
    Top = Bottom + 15
    GroupBoxHeight = 15 * 2 + 20

    ' Create group box
    Set gb = wsInputUI1.OLEObjects.Add(ClassType:="Forms.Image.1", Left:=Left, Top:=Top, Width:=GroupB
oxWidth, Height:=GroupBoxHeight)
    With gb.Object
        .BackColor = RGB(256, 256, 256)
    End With

    AltBoxLabel = "Simulation parameters"
    LabelWidth = Len(AltBoxLabel) * 6
    Set gb = wsInputUI1.OLEObjects.Add(ClassType:="Forms.Label.1", Left:=Left + 10, Top:=Top - 5, Widt
h:=LabelWidth, Height:=15)
    With gb.Object
        .Font.Name = "Courier New"
        .Font.Size = 8
        .TextAlign = fmTextAlignCenter
        .Caption = AltBoxLabel
    End With

    Dim simulationParameters(2) As String
    simulationParameters(1) = "Num Iterations"
    simulationParameters(2) = "Tolerance"
    For i = 1 To UBound(simulationParameters)
        Set cb = wsInputUI1.OLEObjects.Add(ClassType:="Forms.TextBox.1", Left:=Left + 5, Top:=(Top + 1
3) + 15 * (i - 1), Width:=TextBoxWidth, Height:=15)
        cb.Name = simulationParameters(i) & " value"
        With cb.Object
            .Font.Name = "Courier New"
            .Font.Size = 8
        End With

        Set cb = wsInputUI1.OLEObjects.Add(ClassType:="Forms.Label.1", Left:=Left + TextBoxWidth + 10,
 Top:=(Top + 13) + 15 * (i - 1), Width:=LabelWidth, Height:=15)
        With cb.Object
            .Font.Name = "Courier New"
            .Font.Size = 8
            .Caption = simulationParameters(i)
        End With
    Next i

    ' stop the flickering
    Application.ScreenUpdating = True

End Sub


Sub BuildOutputUI()
    Dim i As Integer, j As Integer
    Call setSheets

    ' stop the flickering
    Application.ScreenUpdating = False

    ' clear out UI-1
    Call RemoveAllControls(wsOutputUI1)

    Dim NumAlternatives As Integer
    With wsAlternatives
        NumAlternatives = .Cells(Rows.Count, "A").End(xlUp).row - 1
    End With
    'MsgBox "Number of alternatives: " & NumAlternatives
```

```
BuildUI - 6

    Dim NumTechnologies As Integer
    With wsTechnologies
        NumTechnologies = .Cells(Rows.Count, "A").End(xlUp).row - 1
    End With
    'MsgBox "Number of key technologies: " & NumTechnologies

    Dim NumDurations As Integer
    With wsDurations
        NumDurations = .Cells(Rows.Count, "A").End(xlUp).row - 1
    End With
    'MsgBox "Number of schedule durations: " & NumDurations

    ' This builds the radio button to select the output to display on chart
    '------------------------------------------------------------------
    Dim Left As Double, Right As Double, Top As Double, Bottom As Double
    Dim GroupBoxHeight As Double, GroupBoxWidth As Double, CheckBoxWidth As Double, ComboBoxWidth As D
ouble

    Left = 10
    Top = 10

    GroupBoxHeight = 25
    GroupBoxWidth = 380
    CheckBoxWidth = GroupBoxWidth / 3 - 10

    ' Create group box for radio buttons
    Dim gb As Object
    Set gb = wsOutputUI1.OLEObjects.Add(ClassType:="Forms.Image.1", Left:=Left, Top:=Top, Width:=Group
BoxWidth, Height:=GroupBoxHeight)
    With gb.Object
        .BackColor = RGB(256, 256, 256)
    End With

    Dim AltBoxLabel As Variant
    AltBoxLabel = "Select output"
    Dim LabelWidth As Double
    LabelWidth = Len(AltBoxLabel) * 6
    Set gb = wsOutputUI1.OLEObjects.Add(ClassType:="Forms.Label.1", Left:=Left + 10, Top:=Top - 5, Wid
th:=LabelWidth, Height:=15)
    With gb.Object
        .Font.Name = "Courier New"
        .Font.Size = 8
        .TextAlign = fmTextAlignCenter
        .Caption = AltBoxLabel
    End With

    ' Create radio button
    Dim cb As Object
    Dim outputNames(3) As String
    outputNames(1) = "Performance"
    outputNames(2) = "Schedule"
    outputNames(3) = "Cost"
    For i = 1 To UBound(outputNames)
        Set cb = wsOutputUI1.OLEObjects.Add(ClassType:="Forms.OptionButton.1", Left:=Left + CheckBoxWi
dth * (i - 1) + 25, Top:=Top + 8, Width:=CheckBoxWidth, Height:=15)
        cb.Name = outputNames(i)
        With cb.Object
            .GroupName = "Output Type"
            .Font.Name = "Courier New"
            .Font.Size = 8
            .Caption = outputNames(i)
            .Value = False
        End With
    Next i
    wsOutputUI1.OLEObjects("Performance").Object.Value = True

    ' This builds this alternatiaves to run box
    '------------------------------------------------------------------------
    Left = 400
    Top = 50
```

```
BuildUI - 7

    GroupBoxHeight = 15 * NumAlternatives + 20
    GroupBoxWidth = 205
    CheckBoxWidth = GroupBoxWidth - 10

    ' Create group box for alternatives
    Set gb = wsOutputUI1.OLEObjects.Add(ClassType:="Forms.Image.1", Left:=Left, Top:=Top, Width:=Group
BoxWidth, Height:=GroupBoxHeight)
    With gb.Object
        .BackColor = RGB(256, 256, 256)
    End With

    AltBoxLabel = "Select alternatives"
    LabelWidth = Len(AltBoxLabel) * 6
    Set gb = wsOutputUI1.OLEObjects.Add(ClassType:="Forms.Label.1", Left:=Left + 10, Top:=Top - 5, Wid
th:=LabelWidth, Height:=15)
    With gb.Object
        .Font.Name = "Courier New"
        .Font.Size = 8
        .TextAlign = fmTextAlignCenter
        .Caption = AltBoxLabel
    End With

    ' Create check boxes for alternative
    Dim row As Integer
    row = 2
    For i = 1 To NumAlternatives
        Dim Alternative As String
        Alternative = wsAlternatives.Cells(i + 1, "A").Value

        Set cb = wsOutputUI1.OLEObjects.Add(ClassType:="Forms.CheckBox.1", Left:=Left + 5, Top:=(Top +
13) + 15 * (i - 1), Width:=CheckBoxWidth, Height:=15)
        With cb
            .Name = Alternative
        End With
        With cb.Object
            .GroupName = "Plot Alternative"
            .Font.Name = "Courier New"
            .Font.Size = 8
            .Caption = PadRight(Alternative, 33, " ")
            .Value = False
        End With

        If (Not wsInputUI1.OLEObjects("A: " & i).Object.Value) Then
            cb.Object.Enabled = False
        End If

        row = row + 1
    Next i

    Bottom = Top + GroupBoxHeight

    ' This builds this durations to run box
    '-----------------------------------------------------------------------
    Dim Duration As String

    Left = 400
    Top = Bottom + 15
    GroupBoxHeight = 15 * NumDurations + 20
    GroupBoxWidth = 205
    CheckBoxWidth = 195

    ' Create group box for durations
    Set gb = wsOutputUI1.OLEObjects.Add(ClassType:="Forms.Image.1", Left:=Left, Top:=Top, Width:=Group
BoxWidth, Height:=GroupBoxHeight)
    With gb.Object
        .BackColor = RGB(256, 256, 256)
    End With

    AltBoxLabel = "Select durations"
    LabelWidth = Len(AltBoxLabel) * 6
    Set gb = wsOutputUI1.OLEObjects.Add(ClassType:="Forms.Label.1", Left:=Left + 10, Top:=Top - 5, Wid
```

```
BuildUI - 8
th:=LabelWidth, Height:=15)
    With gb.Object
        .Font.Name = "Courier New"
        .Font.Size = 8
        .TextAlign = fmTextAlignCenter
        .Caption = AltBoxLabel
    End With

    ' Create radio button for durations
    row = 2
    Dim defaultValue As Double
    defaultValue = -99
    For i = 1 To NumDurations
        Duration = wsDurations.Cells(i + 1, "A").Value

        Set cb = wsOutputUI1.OLEObjects.Add(ClassType:="Forms.OptionButton.1", Left:=Left + 5, Top:=(T
op + 13) + 15 * (i - 1), Width:=CheckBoxWidth, Height:=15)
        With cb
            .Name = Duration
        End With
        With cb.Object
            .GroupName = "Durations"
            .Font.Name = "Courier New"
            .Font.Size = 8
            .Caption = Duration & " month schedule duration"
            .Value = False
        End With

        If (Not wsInputUI1.OLEObjects("D: " & i).Object.Value) Then
            cb.Object.Enabled = False
        Else
            If defaultValue = -99 Then
                defaultValue = Duration
            End If
        End If

        row = row + 1
    Next i

    wsOutputUI1.OLEObjects(CStr(defaultValue)).Object.Value = True

    ' Place the blank chart
    '-------------------------------------------------------------------------
    Dim objChart As Object
    For Each objChart In wsOutputUI1.ChartObjects
        objChart.Delete
    Next objChart

    Set objChart = wsOutputUI1.ChartObjects.Add(10, 45, 380, 240)
    objChart.Name = "Output UI-1"


    ' Now I need to place the constraint filters
    '-------------------------------------------------------------------------
    Dim TextBoxWidth As Double

    Left = 10
    Top = Bottom + GroupBoxHeight + 25
    GroupBoxHeight = 15 * 3 + 20
    CheckBoxWidth = 75
    ComboBoxWidth = 85
    TextBoxWidth = 55
    GroupBoxWidth = CheckBoxWidth + ComboBoxWidth + 2 * TextBoxWidth + 55

    ' Create group box
    Set gb = wsOutputUI1.OLEObjects.Add(ClassType:="Forms.Image.1", Left:=Left, Top:=Top, Width:=Group
BoxWidth, Height:=GroupBoxHeight)
    With gb.Object
        .BackColor = RGB(256, 256, 256)
    End With
```

```
BuildUI - 9

    AltBoxLabel = "Constraint filters"
    LabelWidth = Len(AltBoxLabel) * 6
    Set gb = wsOutputUI1.OLEObjects.Add(ClassType:="Forms.Label.1", Left:=Left + 10, Top:=Top - 5, Wid
th:=LabelWidth, Height:=15)
    With gb.Object
        .Font.Name = "Courier New"
        .Font.Size = 8
        .TextAlign = fmTextAlignCenter
        .Caption = AltBoxLabel
    End With

    For i = 1 To UBound(outputNames)
        ' we need a checkbox
        Set cb = wsOutputUI1.OLEObjects.Add(ClassType:="Forms.CheckBox.1", Left:=Left + 5, Top:=(Top +
13) + 15 * (i - 1), Width:=CheckBoxWidth, Height:=15)
        cb.Name = outputNames(i) & " checkbox"
        With cb.Object
            .GroupName = "Constraints"
            .Font.Name = "Courier New"
            .Font.Size = 8
            .Caption = outputNames(i)
            .Value = False
        End With

        ' we need a dropdown
        Set cb = wsOutputUI1.OLEObjects.Add(ClassType:="Forms.ComboBox.1", Left:=Left + CheckBoxWidth
+ 10, Top:=(Top + 13) + 15 * (i - 1), Width:=ComboBoxWidth, Height:=15)
        cb.Name = outputNames(i) & " combobox"
        With cb.Object
            .Font.Name = "Courier New"
            .Font.Size = 8
            .AddItem "is >= than"
            .AddItem "is <= than"
            .AddItem "is between"
        End With

        ' we need two text boxes
        Set cb = wsOutputUI1.OLEObjects.Add(ClassType:="Forms.TextBox.1", Left:=Left + CheckBoxWidth +
ComboBoxWidth + 10, Top:=(Top + 13) + 15 * (i - 1), Width:=TextBoxWidth, Height:=15)
        cb.Name = outputNames(i) & " lowerbound"
        With cb.Object
            .Font.Name = "Courier New"
            .Font.Size = 8
        End With

        LabelWidth = 25
        Set cb = wsOutputUI1.OLEObjects.Add(ClassType:="Forms.Label.1", Left:=Left + CheckBoxWidth + C
omboBoxWidth + TextBoxWidth + 10, Top:=(Top + 13) + 15 * (i - 1), Width:=LabelWidth, Height:=15)
        With cb.Object
            .Font.Name = "Courier New"
            .Font.Size = 8
            .TextAlign = fmTextAlignCenter
            .Caption = "and"
        End With

        Set cb = wsOutputUI1.OLEObjects.Add(ClassType:="Forms.TextBox.1", Left:=Left + CheckBoxWidth +
ComboBoxWidth + TextBoxWidth + LabelWidth + 10, Top:=(Top + 13) + 15 * (i - 1), Width:=TextBoxWidth,
Height:=15)
        cb.Name = outputNames(i) & " upperbound"
        With cb.Object
            .Font.Name = "Courier New"
            .Font.Size = 8
        End With
    Next i

    ' stop the flickering
    Application.ScreenUpdating = True

End Sub
```

```
BuildUI - 10

Sub Build3dUI()
    Dim i As Integer, j As Integer
    Call setSheets

    ' stop the flickering
    Application.ScreenUpdating = False

    ' clear out UI-1
    Call RemoveAllControls(wsOutput3d)

    Dim NumAlternatives As Integer
    With wsAlternatives
        NumAlternatives = .Cells(Rows.Count, "A").End(xlUp).row - 1
    End With
    'MsgBox "Number of alternatives: " & NumAlternatives

    Dim NumTechnologies As Integer
    With wsTechnologies
        NumTechnologies = .Cells(Rows.Count, "A").End(xlUp).row - 1
    End With
    'MsgBox "Number of key technologies: " & NumTechnologies

    Dim NumDurations As Integer
    With wsDurations
        NumDurations = .Cells(Rows.Count, "A").End(xlUp).row - 1
    End With
    'MsgBox "Number of schedule durations: " & NumDurations

    ' This builds the radio button to select the output to display on chart
    '------------------------------------------------------------------------
    Dim Left As Double, Right As Double, Top As Double, Bottom As Double
    Dim GroupBoxHeight As Double, GroupBoxWidth As Double, CheckBoxWidth As Double, ComboBoxWidth As D
ouble

    Left = 10
    Top = 10

    GroupBoxHeight = 25
    GroupBoxWidth = 380
    CheckBoxWidth = GroupBoxWidth / 3 - 10

    ' Create group box for radio buttons
    Dim gb As Object
    Set gb = wsOutput3d.OLEObjects.Add(ClassType:="Forms.Image.1", Left:=Left, Top:=Top, Width:=GroupB
oxWidth, Height:=GroupBoxHeight)
    With gb.Object
        .BackColor = RGB(256, 256, 256)
    End With

    ' This builds this alternatiaves to run box
    '------------------------------------------------------------------------
    Left = 400
    Top = 50

    GroupBoxHeight = 15 * NumAlternatives + 20
    GroupBoxWidth = 205
    CheckBoxWidth = GroupBoxWidth - 10

    ' Create group box for alternatives
    Set gb = wsOutput3d.OLEObjects.Add(ClassType:="Forms.Image.1", Left:=Left, Top:=Top, Width:=GroupB
oxWidth, Height:=GroupBoxHeight)
    With gb.Object
        .BackColor = RGB(256, 256, 256)
    End With

    Dim AltBoxLabel As Variant
    AltBoxLabel = "Select alternatives"
    Dim LabelWidth As Double
    LabelWidth = Len(AltBoxLabel) * 6
    Set gb = wsOutput3d.OLEObjects.Add(ClassType:="Forms.Label.1", Left:=Left + 10, Top:=Top - 5, Widt
h:=LabelWidth, Height:=15)
```

```
BuildUI - 11

    With gb.Object
        .Font.Name = "Courier New"
        .Font.Size = 8
        .TextAlign = fmTextAlignCenter
        .Caption = AltBoxLabel
    End With

    ' Create check boxes for alternative
    Dim row As Integer
    row = 2
    For i = 1 To NumAlternatives
        Dim Alternative As String
        Alternative = wsAlternatives.Cells(i + 1, "A").Value

        Dim cb As Object
        Set cb = wsOutput3d.OLEObjects.Add(ClassType:="Forms.CheckBox.1", Left:=Left + 5, Top:=(Top +
13) + 15 * (i - 1), Width:=CheckBoxWidth, Height:=15)
        With cb
            .Name = Alternative
        End With
        With cb.Object
            .GroupName = "Plot Alternative"
            .Font.Name = "Courier New"
            .Font.Size = 8
            .Caption = PadRight(Alternative, 33, " ")
            .Value = False
        End With

        If (Not wsInputUI1.OLEObjects("A: " & i).Object.Value) Then
            cb.Object.Enabled = False
        End If

        row = row + 1
    Next i

    Bottom = Top + GroupBoxHeight

    ' This builds this durations to run box
    '--------------------------------------------------------------------------
    Dim Duration As String

    Left = 400
    Top = Bottom + 15
    GroupBoxHeight = 15 * NumDurations + 20
    GroupBoxWidth = 205
    CheckBoxWidth = 195

    ' Create group box for durations
    Set gb = wsOutput3d.OLEObjects.Add(ClassType:="Forms.Image.1", Left:=Left, Top:=Top, Width:=GroupB
oxWidth, Height:=GroupBoxHeight)
    With gb.Object
        .BackColor = RGB(256, 256, 256)
    End With

    AltBoxLabel = "Select durations"
    LabelWidth = Len(AltBoxLabel) * 6
    Set gb = wsOutput3d.OLEObjects.Add(ClassType:="Forms.Label.1", Left:=Left + 10, Top:=Top - 5, Widt
h:=LabelWidth, Height:=15)
    With gb.Object
        .Font.Name = "Courier New"
        .Font.Size = 8
        .TextAlign = fmTextAlignCenter
        .Caption = AltBoxLabel
    End With

    ' Create radio button for durations
    row = 2
    Dim defaultValue As Double
    defaultValue = -99
    For i = 1 To NumDurations
        Duration = wsDurations.Cells(i + 1, "A").Value
```

```
BuildUI - 12

        Set cb = wsOutput3d.OLEObjects.Add(ClassType:="Forms.OptionButton.1", Left:=Left + 5, Top:=(To
p + 13) + 15 * (i - 1), Width:=CheckBoxWidth, Height:=15)
        With cb
            .Name = Duration
        End With
        With cb.Object
            .GroupName = "Durations Bubble"
            .Font.Name = "Courier New"
            .Font.Size = 8
            .Caption = Duration & " month schedule duration"
            .Value = False
        End With

        If (Not wsInputUI1.OLEObjects("D: " & i).Object.Value) Then
            cb.Object.Enabled = False
        Else
            If defaultValue = -99 Then
                defaultValue = Duration
            End If
        End If

        row = row + 1
    Next i

    wsOutput3d.OLEObjects(CStr(defaultValue)).Object.Value = True

    ' Place the blank chart
    '--------------------------------------------------------------------
    Dim objChart As Object
    For Each objChart In wsOutput3d.ChartObjects
        objChart.Delete
    Next objChart

    Set objChart = wsOutput3d.ChartObjects.Add(10, 45, 380, 240)
    objChart.Name = "Output 3d"

    ' stop the flickering
    Application.ScreenUpdating = True

End Sub
```

DataViz Module

The DataViz module creates the output from the model runs.

```
DataViz - 1

Option Explicit

' clear out old chart objects
Sub DeleteAllCharts(ws As Worksheet)
    Call setSheets
    Dim objChart As ChartObject
    For Each objChart In ws.ChartObjects
        objChart.Delete
    Next
End Sub

' re-create the output chart
Sub CreateChart(ws As Worksheet, chartName As String)
    Call setSheets
    Dim objChart As Object
    Set objChart = ws.ChartObjects.Add(10, 45, 380, 240)
    objChart.Name = chartName
End Sub

' remove all series from given chart
Sub RemoveAllSeries(chartName As String)
    Dim objChart As ChartObject
    Set objChart = wsOutputUI1.ChartObjects(chartName)
    Dim s As Series
    For Each s In objChart.chart.SeriesCollection
        s.Delete
    Next s
End Sub

' fix the legend so it doesn't display series with no data
Sub FixLegend(chartName As String)
    Dim i As Integer
    Dim objChart As ChartObject
    Set objChart = wsOutputUI1.ChartObjects(chartName)
    For i = objChart.chart.SeriesCollection.Count To 1 Step -1
        If UBound(objChart.chart.SeriesCollection(i).Values) = 0 Then
            Dim j As Integer
            j = objChart.chart.Legend.LegendEntries.Count
            objChart.chart.Legend.LegendEntries(i).Delete
        End If
    Next i
End Sub

Function getOption(ws As Worksheet, gName As String) As String
    getOption = "ERROR"
    Dim oleObj As Object
    For Each oleObj In ws.OLEObjects
        If oleObj.progID = "Forms.OptionButton.1" Then
            If oleObj.Object.GroupName = gName Then
                If oleObj.Object.Value Then
                    getOption = oleObj.Name
                End If
            End If
        End If
    Next oleObj
End Function

Function filteredObs(ws As Worksheet, filter As String, obsIndex As Long) As Boolean
    filteredObs = False

    If ws.OLEObjects(filter & " checkbox").Object.Value Then
        Dim conditional As String
        conditional = ws.OLEObjects(filter & " combobox").Object.Value

        Dim lowerbound As Double
        If Not ws.OLEObjects(filter & " lowerbound").Object.Value = "" Then
            lowerbound = ws.OLEObjects(filter & " lowerbound").Object.Value
        Else
            MsgBox "You must specify a lowerbound for " & filter
            End
        End If
```

DataViz - 2

```
        ' what data are we filtering on?
        Dim colNum As Integer
        If filter = "Performance" Then
            colNum = 4
        ElseIf filter = "Schedule" Then
            colNum = 5
        Else
            colNum = 6
        End If

        Dim compValue As Double
        compValue = wsOutputConsequence.Cells(obsIndex, colNum)

        lowerbound = ws.OLEObjects(filter & " lowerbound").Object.Value
        If conditional = "is <= than" Then
            If compValue >= lowerbound Then
                filteredObs = True
            End If
        ElseIf conditional = "is >= than" Then
            If compValue <= lowerbound Then
                filteredObs = True
            End If
        Else
            Dim upperbound As Double
            If Not ws.OLEObjects(filter & " upperbound").Object.Value = "" Then
                upperbound = ws.OLEObjects(filter & " upperbound").Object.Value
            Else
                MsgBox "You must specify a upperbound for " & filter
                End
            End If
            If compValue < lowerbound Or compValue > upperbound Then
                filteredObs = True
            End If
        End If
    End If
End Function

Function getSeries(ws As Worksheet, seriesNames As Collection, colNum As Integer, simDur As Double) As
 Collection
    Dim i As Long
    Set getSeries = New Collection

    Dim seriesName As Variant
    Dim seriesObs As Collection
    Dim removeObs As Collection
    Set removeObs = New Collection
    For Each seriesName In seriesNames
        Dim cs As ChartSeries
        Set cs = New ChartSeries
        With cs
            .seriesName = seriesName
            Set .SeriesData = New Collection
        End With

        For i = 2 To wsOutputConsequence.Cells(Rows.Count, "A").End(xlUp).row
            If seriesName = wsOutputConsequence.Cells(i, 1) Then
                If simDur = wsOutputConsequence.Cells(i, 2) Then
                    Dim bPerf As Boolean, bSched As Boolean, bCost As Boolean
                    bPerf = Not filteredObs(ws, "Performance", i)
                    bSched = Not filteredObs(ws, "Schedule", i)
                    bCost = Not filteredObs(ws, "Cost", i)

                    If bPerf And bSched And bCost Then
                        cs.SeriesData.Add wsOutputConsequence.Cells(i, colNum)
                    End If
                End If
            End If
        Next i
        'MsgBox "The series " & cs.seriesName & " has " & cs.seriesData.Count & " observations"
```

```
DataViz - 3
        If (cs.SeriesData.Count > 0) Then
            getSeries.Add cs
        Else
            MsgBox "Constraints have filtered out all data points for " & seriesName
        End If
    Next seriesName

End Function


Sub MainOutputRoutine()
    Dim i As Integer, j As Integer, k As Integer
    Dim oleObj As Object

    Dim chartName As String
    chartName = "Output UI-1"

    Call setSheets
    Call DeleteAllCharts(wsOutputUI1)
    Call CreateChart(wsOutputUI1, chartName)

    Dim objChart As Object
    Set objChart = wsOutputUI1.ChartObjects(chartName)

    ' get x-axis variable
    '-----------------------------------------------------------------------
    Dim outputType As String
    outputType = getOption(wsOutputUI1, "Output Type")
    'MsgBox outputType

    Dim colNumber As Integer
    If outputType = "Performance" Then
        colNumber = 4
    ElseIf outputType = "Schedule" Then
        colNumber = 5
    ElseIf outputType = "Cost" Then
        colNumber = 6
    End If

    ' get series to plot
    '-----------------------------------------------------------------------
    Dim CheckedAlternatives As New Collection
    Dim AllAlternatives() As String
    For Each oleObj In wsOutputUI1.OLEObjects
        If oleObj.progID = "Forms.CheckBox.1" Then
            If oleObj.Object.GroupName = "Plot Alternative" Then
                'MsgBox OleObj.Name & " is checked: " & OleObj.Object.Value
                If (oleObj.Object.Value) Then
                    CheckedAlternatives.Add oleObj.Name
                End If
                j = j + 1
                ReDim Preserve AllAlternatives(1 To j)
                AllAlternatives(j) = oleObj.Name
            End If
        End If
    Next oleObj

    ' ERROR checking
    If (CheckedAlternatives.Count = 0) Then
        MsgBox "You must select an alternative"
        End
    End If

    ' filter series by duration
    '-----------------------------------------------------------------------
    Dim simDuration As String
    simDuration = getOption(wsOutputUI1, "Durations")
    'MsgBox simDuration

    ' get the data from the output sheet
    '-----------------------------------------------------------------------
```

DataViz - 4

```
Dim ChartData As Collection
Set ChartData = getSeries(wsOutputUI1, CheckedAlternatives, colNumber, CDbl(simDuration))

' ERROR checking: do we have any data to plot
'------------------------------------------------------------------------
If (ChartData.Count = 0) Then
    MsgBox "There is no data to chart!"
    End
End If

' create and empty series for each alternative. do this so colors don't change
'------------------------------------------------------------------------
For i = 1 To UBound(AllAlternatives)
    With objChart.chart
        .SeriesCollection.NewSeries
        With .SeriesCollection(.SeriesCollection.Count)
            .Name = AllAlternatives(i)
        End With
    End With
Next i

' add the data to the chart
'------------------------------------------------------------------------
If outputType = "Performance" Then
    Call CreateHistogram(objChart, ChartData)
ElseIf outputType = "Schedule" Then
    Call CreateCDF(objChart, ChartData)
Else
    Call CreateCDF(objChart, ChartData, 100)
End If

' remove non-visible series from chart legend
Call FixLegend(chartName)

If (outputType = "Performance") Then
    outputType = "Performance degradation"
End If

With objChart.chart.Axes(xlCategory)
    .HasTitle = True
    .AxisTitle.text = outputType
End With
End Sub

Sub CreateHistogram(objChart As Object, ChartData As Collection)
Dim i As Integer
Dim intBins As Integer
Dim dblBinWidth As Double
Dim X() As Double, Y() As Double

' performance is a value from 0 to 5
intBins = 6
dblBinWidth = 1

ReDim breaks(1 To intBins) As Double
ReDim freq(1 To intBins) As Double

'Linear interpolation
For i = 1 To intBins
    breaks(i) = i
Next i

' set some chart "global" values
'------------------------------------------------------------------------
With objChart.chart
    .chartType = xlColumnClustered
    .SetElement (msoElementPrimaryValueGridLinesNone) 'turns off gridlines

    With .Axes(xlValue)
        .HasTitle = True
        .AxisTitle.text = "Probability"
```

```
DataViz - 5
                .MinimumScale = 0
                .MaximumScale = 1
                .CrossesAt = 0
                .MajorUnit = 0.2
                .MinorUnit = 0.1
            End With
        End With

        ' loop over series: compute histogram, i.e., bin data, and add to chart
        '----------------------------------------------------------------------
        Dim cs As ChartSeries
        For Each cs In ChartData
            Dim seriesName As String
            seriesName = cs.seriesName

            'Assign initial value for the frequency array
            For i = 1 To intBins
                freq(i) = 0
            Next i

            'Counting the number of occurrences for each of the bins
            Dim chartOb As Variant
            For Each chartOb In cs.SeriesData
                If (chartOb < breaks(1)) Then
                    freq(1) = freq(1) + 1
                End If
                If (chartOb >= breaks(intBins - 1)) Then
                    freq(intBins) = freq(intBins) + 1
                End If
                For i = 2 To intBins - 1
                    If (chartOb >= breaks(i - 1) And chartOb < breaks(i)) Then
                        freq(i) = freq(i) + 1
                    End If
                Next i
            Next chartOb

            ReDim X(1 To intBins)
            ReDim Y(1 To intBins)
            For i = 1 To intBins
                'MsgBox "bin = " & i & ", freq = " & freq(i)
                X(i) = i - 1
                Y(i) = freq(i) / cs.SeriesData.Count
            Next i

            With objChart.chart
                Dim s As Series
                For Each s In .SeriesCollection
                    If s.Name = seriesName Then
                        s.Values = Y
                        s.XValues = X
                        's.ApplyDataLabels
                        's.DataLabels.NumberFormat = "0.0%"
                    End If
                Next s
            End With
        Next cs
    End Sub

    Sub CreateCDF(objChart As Object, ChartData As Collection, Optional steps As Integer)
        Dim i As Integer
        Dim cs As ChartSeries
        Dim X As New Collection, Y As New Collection

        ' set some chart "global" values
        '----------------------------------------------------------------------
        With objChart.chart
            .chartType = xlXYScatterLines
            .SetElement (msoElementPrimaryValueGridLinesNone) 'turns off gridlines

            With .Axes(xlValue)
                .HasTitle = True
```

```
DataViz - 6
            .AxisTitle.text = "Cumulative Probability"
            .MinimumScale = 0
            .MaximumScale = 1
            .CrossesAt = 0
            .MajorUnit = 0.2
            .MinorUnit = 0.1
        End With
    End With

    ' compute the min/max so we can re-scale the x-axis
    '-------------------------------------------------------------------------
    Dim dataPoints As New Collection
    For Each cs In ChartData
        Dim dataPoint As Variant
        For Each dataPoint In cs.SeriesData
            dataPoints.Add dataPoint
        Next dataPoint
    Next cs

    Dim min As Double
    min = Application.min(collectionToArray(dataPoints))
    Dim max As Double
    max = Application.max(collectionToArray(dataPoints))

    With objChart.chart.Axes(xlCategory)
        .MinimumScale = min * 0.95
        .MaximumScale = max * 1.05
        .TickLabels.NumberFormat = "#,##0"
    End With

    ' loop over series: compute histogram, i.e., bin data, and add to chart
    '-------------------------------------------------------------------------
    For Each cs In ChartData
        Dim seriesName As String
        seriesName = cs.seriesName

        Dim SeriesData() As Variant
        SeriesData = collectionToArray(cs.SeriesData)

        Set X = New Collection
        Set Y = New Collection

        ' compute the empirical CDF
        max = Application.max(SeriesData)
        If steps = 0 Then
            min = Application.min(SeriesData)
            For i = min To max
                ' sometimes the CDF has only a single point
                If min = max Then
                    X.Add min, key:="Zero"
                    Y.Add 0, key:="Zero"
                End If
                X.Add i, key:=CStr(i)
                Y.Add empiricalCDF(SeriesData, X.Item(CStr(i)))
            Next i
        Else
            For i = 0 To steps
                X.Add i / steps * max, key:=CStr(i)
                Y.Add empiricalCDF(SeriesData, X.Item(CStr(i)))
                'Debug.Print "x: " & X(i)
                'Debug.Print "y: " & Y(i)
            Next i
        End If

        With objChart.chart
            Dim s As Series
            For Each s In .SeriesCollection
                If s.Name = seriesName Then
                    s.Values = collectionToArray(Y)
                    s.XValues = collectionToArray(X)
                End If
```

```
DataViz - 7
                s.MarkerStyle = xlMarkerStyleNone
            Next s
        End With
    Next cs
End Sub

Sub Create3dChart()
    Dim i As Integer, j As Integer, k As Integer
    Dim oleObj As Object

    Dim chartName As String
    chartName = "Output 3d"

    Call setSheets
    Call DeleteAllCharts(wsOutput3d)
    Call CreateChart(wsOutput3d, chartName)

    Dim objChart As Object
    Set objChart = wsOutput3d.ChartObjects(chartName)

    ' get series to plot
    '--------------------------------------------------------------------
    Dim CheckedAlternatives As New Collection
    Dim AllAlternatives() As String
    For Each oleObj In wsOutput3d.OLEObjects
        If oleObj.progID = "Forms.CheckBox.1" Then
            If oleObj.Object.GroupName = "Plot Alternative" Then
                'MsgBox oleObj.Name & " is checked: " & oleObj.Object.Value
                If (oleObj.Object.Value) Then
                    CheckedAlternatives.Add oleObj.Name
                End If
                j = j + 1
                ReDim Preserve AllAlternatives(1 To j)
                AllAlternatives(j) = oleObj.Name
            End If
        End If
    Next oleObj

    ' ERROR checking
    If (CheckedAlternatives.Count = 0) Then
        MsgBox "You must select an alternative"
        End
    End If

    ' filter series by duration
    '--------------------------------------------------------------------
    Dim simDuration As String
    simDuration = getOption(wsOutput3d, "Durations Bubble")
    'MsgBox simDuration

    ' get the data from the output sheet
    '--------------------------------------------------------------------
    Dim c As Collection, s As Collection, p As Collection
    Set c = getSeries(wsOutputUI1, CheckedAlternatives, 6, CDbl(simDuration))
    Set s = getSeries(wsOutputUI1, CheckedAlternatives, 5, CDbl(simDuration))
    Set p = getSeries(wsOutputUI1, CheckedAlternatives, 4, CDbl(simDuration))

    ' get the min, max, mean of each variable
    Dim cs As ChartSeries

    Dim cAve As Collection
    Set cAve = New Collection
    For Each cs In c
        cAve.Add Application.Average(collectionToArray(cs.SeriesData))
    Next cs

    Dim cPlus As New Collection
    Dim cMinus As New Collection
    For Each cs In c
        cPlus.Add Application.WorksheetFunction.Percentile(collectionToArray(cs.SeriesData), 0.95)
        cMinus.Add Application.WorksheetFunction.Percentile(collectionToArray(cs.SeriesData), 0.05)
```

```
DataViz - 8

    Next cs

    Dim sAve As Collection
    Set sAve = New Collection
    For Each cs In s
        sAve.Add Application.Average(collectionToArray(cs.SeriesData))
    Next cs

    Dim pAve As Collection
    Set pAve = New Collection
    For Each cs In p
        pAve.Add Application.Average(collectionToArray(cs.SeriesData))
    Next cs

    ' add the data to the chart
    '------------------------------------------------------------------------
    With objChart.chart
        .chartType = xlXYScatter
        For i = 1 To CheckedAlternatives.Count
            With .SeriesCollection.NewSeries
                .Name = CheckedAlternatives(i)
                .XValues = pAve.Item(i)
                .Values = cAve.Item(i)

                .Format.Fill.Solid
                .Format.Fill.Transparency = 0.25

                ' this works because we know there is only one point in each series
                .Points(1).HasDataLabel = True
                .Points(1).DataLabel.text = CheckedAlternatives(i) & " (Schedule =  " & Round(sAve.Ite
m(i), 1) & ")"
            End With

            ' make everything a circle
            .SeriesCollection(i).MarkerStyle = xlMarkerStyleCircle
        Next i
    End With

    ' set some chart "global" values
    '------------------------------------------------------------------------
    With objChart.chart
        .SetElement (msoElementLegendNone) ' turns off legend
        .SetElement (msoElementPrimaryValueGridLinesNone) 'turns off gridlines
        .SetElement (msoElementErrorBarPercentage) ' add the error bars to the chart

        With .Axes(xlCategory)
            .HasTitle = True
            .AxisTitle.text = "Performance Degradation (0-5 scale)"
            '.TickLabels.NumberFormat = "0%"
            '.TickLabels.Orientation = 90
            .MinimumScale = -1
            .MaximumScale = 5.5
            .CrossesAt = 1
        End With

        With .Axes(xlValue)
            .HasTitle = True
            .AxisTitle.text = "RTD&E Cost ($M)"
            '.TickLabels.NumberFormat = "0%"
            '.TickLabels.Orientation = 90
            '.MinimumScale = 0
            '.MaximumScale = 5
            .CrossesAt = 0
        End With
    End With
End Sub

Sub Build3dChart(seriesName() As String, X() As Double, Y() As Double, Z() As Double)
    Dim objChart As Object
    Set objChart = wsOutput3d.ChartObjects("Output 3d")
    With objChart.chart
```

```
DataViz - 9

        .chartType = xlBubble
        .HasLegend = False

        With .SeriesCollection.NewSeries
            .Name = seriesName
            .XValues = X
            .Values = Y
            .BubbleSizes = Z
        End With

        Dim i As Integer
        For i = 1 To .SeriesCollection(1).Points.Count
            With .SeriesCollection(1).Points(i)
                .HasDataLabel = True
                With .DataLabel
                    .text = Z(i)
                End With
            End With
        Next i
    End With
End Sub

Sub CraigsOldChartingCode()

    ' create some histograms
    '-----------------------------------------------------------------------
    Dim intBins As Integer
    Dim dblBinWidth As Double

    ThisWorkbook.Sheets("histograms").ChartObjects.Delete 'will want these eventually
    ThisWorkbook.Sheets("histograms").ScrollBars.Delete

    ' PERFORMANCE
    '++++++++++++++++++++++++++++++++++++++++++++++++++++++++++++++++++++++++++

    ' performance is a value from 0 to 5
    intBins = 6
    dblBinWidth = 1

    ReDim breaks(1 To intBins) As Double
    ReDim freq(1 To intBins) As Double

    'Sort the array
    Call QuickSort(dblPerf, LBound(dblPerf), UBound(dblPerf))

    'Assign initial value for the frequency array
    For i = 1 To intBins
        freq(i) = 0
    Next i

    'Linear interpolation
    For i = 1 To intBins
        breaks(i) = i
    Next i

    'Counting the number of occurrences for each of the bins
    For i = 1 To UBound(dblPerf)
        If (dblPerf(i) < breaks(1)) Then freq(1) = freq(1) + 1
        If (dblPerf(i) >= breaks(intBins - 1)) Then freq(intBins) = freq(intBins) + 1
        For j = 2 To intBins - 1
            If (dblPerf(i) >= breaks(j - 1) And dblPerf(i) < breaks(j)) Then freq(j) = freq(j) + 1
        Next j
    Next i

    'Display the frequency distribution on the active worksheet
    ThisWorkbook.Sheets("histograms").Cells(1, 1) = "bins"
    ThisWorkbook.Sheets("histograms").Cells(1, 2) = "perf"
    For i = 1 To intBins
        ThisWorkbook.Sheets("histograms").Cells(i + 1, 1) = breaks(i) - 1
        ThisWorkbook.Sheets("histograms").Cells(i + 1, 2) = freq(i) / intMCIterations '[cab] put into
proportions
```

```
DataViz - 10

    Next i

    '[cab] Generate  Chart

    Dim PerfChart As ChartObject
    Set PerfChart = ThisWorkbook.Sheets("histograms").ChartObjects.Add(1420, 5, 400, 300) 'left, top,
width, height relative to A1
    With PerfChart.chart
        .SetSourceData Source:=Range(ThisWorkbook.Sheets("histograms").Cells(2, 2), ThisWorkbook.Sheet
s("histograms").Cells(intBins + 1, 2)) 'get the two series
        With .SeriesCollection(1)
            .XValues = Range(ThisWorkbook.Sheets("histograms").Cells(2, 1), ThisWorkbook.Sheets("histo
grams").Cells(intBins + 1, 1))
            .Format.Line.ForeColor.ObjectThemeColor = msoThemeColorAccent2
            .Format.Fill.ForeColor.ObjectThemeColor = msoThemeColorAccent2
            .ApplyDataLabels
            .DataLabels.NumberFormat = "0.0%"
        End With
        .chartType = xlColumnClustered
        .HasTitle = True
        .HasLegend = False
        .ChartTitle.Caption = "Performance Degradation Distribution (pdf)"
        .SetElement (msoElementPrimaryValueGridLinesNone) 'turns off gridlines
        With .Axes(xlCategory)
            .HasTitle = True
            .AxisTitle.text = "Performance Consequence (greater is worse)"
        End With
        With .Axes(xlValue)
            .HasTitle = True
            .AxisTitle.text = "Probability"
            .MinimumScale = 0
            .MaximumScale = 1
            .CrossesAt = 0
            .MajorUnit = 0.2
            .MinorUnit = 0.1
        End With
    End With
    ThisWorkbook.Sheets("histograms").Cells(intBins + 2, 1) = "bins"
    ThisWorkbook.Sheets("histograms").Cells(intBins + 2, 2) = "perf cdf"
    For i = 1 To intBins
        ThisWorkbook.Sheets("histograms").Cells(intBins + 2 + i, 1) = ThisWorkbook.Sheets("histograms"
).Cells(i + 1, 1)
        If i = 1 Then
            ThisWorkbook.Sheets("histograms").Cells(intBins + 2 + i, 2) = ThisWorkbook.Sheets("histogr
ams").Cells(i + 1, 2)
        Else
            ThisWorkbook.Sheets("histograms").Cells(intBins + 2 + i, 2) = ThisWorkbook.Sheets("histogr
ams").Cells(intBins + 1 + i, 2) + ThisWorkbook.Sheets("histograms").Cells(i + 1, 2)
        End If
    Next i


    ' SCHEDULE
    '+++++++++++++++++++++++++++++++++++++++++++++++++++++++++++++++++++++++++++

    ' schedule is a value from 1 to rt
    intBins = CInt(intEndOfTime)
    dblBinWidth = 1

    ReDim breaks(1 To intBins) As Double
    ReDim freq(1 To intBins) As Double

    'Sort the array
    Call QuickSort(dblSched, LBound(dblSched), UBound(dblSched))

    'Assign initial value for the frequency array
    For i = 1 To intBins
        freq(i) = 0
    Next i
```

```
DataViz - 11

    'Linear interpolation
    For i = 1 To intBins
        breaks(i) = i
    Next i

    'Counting the number of occurrences for each of the bins
    For i = 1 To UBound(dblSched)
        If (dblSched(i) <= breaks(1)) Then freq(1) = freq(1) + 1
        If (dblSched(i) >= breaks(intBins - 1)) Then freq(intBins) = freq(intBins) + 1
        For j = 2 To intBins - 1
            If (dblSched(i) > breaks(j - 1) And dblSched(i) <= breaks(j)) Then freq(j) = freq(j) + 1
        Next j
    Next i

    'Display the frequency distribution on the active worksheet
    ThisWorkbook.Sheets("histograms").Cells(1, 3) = "bins"
    ThisWorkbook.Sheets("histograms").Cells(1, 4) = "sched"
    ThisWorkbook.Sheets("histograms").Cells(1, 5) = "cdf" '[cab] print cdf as well

    For i = 1 To intBins
        ThisWorkbook.Sheets("histograms").Cells(i + 1, 3) = breaks(i)
        ThisWorkbook.Sheets("histograms").Cells(i + 1, 4) = freq(i) / intMCIterations '[cab] put into
proportions
            If i = 1 Then
                ThisWorkbook.Sheets("histograms").Cells(i + 1, 5) = ThisWorkbook.Sheets("histograms").
Cells(i + 1, 4)
            Else
                ThisWorkbook.Sheets("histograms").Cells(i + 1, 5) = ThisWorkbook.Sheets("histograms").
Cells(i, 5) + ThisWorkbook.Sheets("histograms").Cells(i + 1, 4) '[cab] cdf for schedule
            End If
    Next i

    '[cab] Generate Interactive Chart

    For i = 1 To intBins '[cab] find first bin with positive density
        If freq(i) <> 0 Then
            Exit For
        End If
    Next i

    intSchedmin = i

    For i = intBins To intSchedmin Step -1 '[cab] find last bin with positive density
        If freq(i) <> 0 Then
            Exit For
        End If
    Next i
    intSchedmax = i


    Dim SampleChart As ChartObject
    Set SampleChart = ThisWorkbook.Sheets("histograms").ChartObjects.Add(600, 5, 400, 300) 'left, top,
width, height relative to A1
    With SampleChart.chart
        .SetSourceData Source:=Range(ThisWorkbook.Sheets("histograms").Cells(intSchedmin, 4), ThisWork
book.Sheets("histograms").Cells(intSchedmax + 2, 5)) 'get the two series
        .SeriesCollection(1).XValues = Range(ThisWorkbook.Sheets("histograms").Cells(intSchedmin, 3),
(ThisWorkbook.Sheets("histograms").Cells(intSchedmax + 2, 3)))
        .SeriesCollection(2).XValues = Range(ThisWorkbook.Sheets("histograms").Cells(intSchedmin, 3),
(ThisWorkbook.Sheets("histograms").Cells(intSchedmax + 2, 3)))

        .chartType = xlXYScatter 'start with scatter chart
        .HasTitle = True
        .ChartTitle.Caption = "Schedule Distribution (red=cdf, blue-pdf)"
        .Axes(xlCategory).HasTitle = True
        .Axes(xlCategory).AxisTitle.text = "Months"
        .Axes(xlValue).HasTitle = True
        .Axes(xlValue).AxisTitle.text = "Probability"
    '    .Axes(xlSecondary).HasTitle = True
    '    .Axes(xlSecondary).AxisTitle.Text = "Secondary"
        .HasLegend = False
```

```
DataViz - 12
        .Axes(xlCategory).TickLabelSpacing = 10 'these can be handy for chart scaling in applications
        .Axes(xlValue).MinimumScale = 0
        .Axes(xlValue).MaximumScale = 1
        .Axes(xlValue).CrossesAt = 0
        .Axes(xlValue).MajorUnit = 0.2
        .Axes(xlValue).MinorUnit = 0.1
        .SetElement (msoElementPrimaryValueGridLinesNone) 'turns off gridlines
    End With
    'Put data readout beside the chart
    ThisWorkbook.Sheets("histograms").Cells(22, 15) = "X Value"
    ThisWorkbook.Sheets("histograms").Cells(22, 16) = "Cum. Prob."
    ThisWorkbook.Sheets("histograms").Cells(22, 17) = "Frequency"
    ThisWorkbook.Sheets("histograms").Cells(23, 15) = intSchedmin
    Application.ThisWorkbook.Sheets("histograms").Range("p23").Formula = "=Lookup(o23,c2:c10000,e2:e10
00)" 'actually write formula for dynamic updating
    Application.ThisWorkbook.Sheets("histograms").Range("q23").Formula = "=Lookup(o23,c2:c1000,d2:d100
0)"

    With ThisWorkbook.Sheets("histograms").Range("p23:q23")
        .Style = "Percent"
        .NumberFormat = "0.0%"
    End With

    Dim Scrollb As ScrollBar ' NOTE microsoft library has all properties
    Set Scrollb = ThisWorkbook.Sheets("histograms").ScrollBars.Add(600, 305, 400, 10) 'adds the declar
ed scroll bar at those points in sheet
    With Scrollb 'scroll bar properties
        .Value = intSchedmin
        .min = intSchedmin
        .max = intSchedmax
        .SmallChange = 1
        .LargeChange = 10
        .LinkedCell = "=histograms!$o$23"
        .Display3DShading = True
    End With

    With SampleChart.chart
        .SeriesCollection.NewSeries 'declares a new series
        .SeriesCollection(3).chartType = xlXYScatter 'makes it a scatterplot
        .SeriesCollection(3).XValues = "=histograms!$o$23" 'x value, linked to scroll bar
        .SeriesCollection(3).Values = "=histograms!$p$23" 'y value'
        .Axes(xlCategory).MinimumScale = intSchedmin
        .Axes(xlCategory).MaximumScale = intSchedmax + 1
        .SeriesCollection(3).MarkerStyle = -4168
        .SeriesCollection(3).Format.Fill.Visible = msoFalse
        .SeriesCollection(3).Format.Line.Visible = msoTrue
        .SeriesCollection(3).Format.Line.ForeColor.ObjectThemeColor = msoThemeColorText1
        .SeriesCollection(1).Format.Line.Visible = msoTrue 'formats the pdf series
        .SeriesCollection(2).Format.Line.Visible = msoTrue 'formats the pdf series
        .SeriesCollection(1).Format.Line.ForeColor.ObjectThemeColor = msoThemeColorAccent1
        .SeriesCollection(1).Format.Line.DashStyle = msoLineSysDot
        .SeriesCollection(1).MarkerStyle = -4142
        .SeriesCollection(2).MarkerStyle = -4142
        .SeriesCollection(2).Format.Line.ForeColor.ObjectThemeColor = msoThemeColorAccent2
    End With

    '[cab] Add Schedule Duration Time

    With SampleChart.chart
        .SeriesCollection.NewSeries 'declares a new series
        With .SeriesCollection(4)
            .XValues = Array(dblReqTime, dblReqTime)
            .Values = Array(0, 1)
            .Format.Line.Visible = msoTrue
            .Format.Line.DashStyle = msoLineDash
        End With
    End With

    ' COST
    '++++++++++++++++++++++++++++++++++++++++++++++++++++++++++++++++++++++++++++++
```

```
DataViz - 13

    intBins = 30

    ReDim breaks(1 To intBins) As Double
    ReDim freq(1 To intBins) As Double

    'Sort the array
    Call QuickSort(dblCost, LBound(dblCost), UBound(dblCost))

    'Assign initial value for the frequency array
    For i = 1 To intBins
        freq(i) = 0
    Next i

    'Linear interpolation
    dblBinWidth = (dblCost(UBound(dblCost)) - dblCost(1)) / intBins
    For i = 1 To intBins
        breaks(i) = dblCost(1) + i * dblBinWidth
    Next i

    'Counting the number of occurrences for each of the bins
    For i = 1 To UBound(dblCost)
        If (dblCost(i) <= breaks(1)) Then freq(1) = freq(1) + 1
        If (dblCost(i) >= breaks(intBins - 1)) Then freq(intBins) = freq(intBins) + 1
        For j = 2 To intBins - 1
            If (dblCost(i) > breaks(j - 1) And dblCost(i) <= breaks(j)) Then freq(j) = freq(j) + 1
        Next j
    Next i

    'Display the frequency distribution on the active worksheet
    ThisWorkbook.Sheets("histograms").Cells(1, 6) = "bins"
    ThisWorkbook.Sheets("histograms").Cells(1, 7) = "cost"
    ThisWorkbook.Sheets("histograms").Cells(1, 8) = "cdf" '[cab] print cdf as well

    For i = 1 To intBins
        'ThisWorkbook.Sheets("histograms").Cells(i + 1, 6) = "(" & Format(breaks(i), "#,##0.00") & ",
" & Format(breaks(i), "#,##0.00") & "]"
        ThisWorkbook.Sheets("histograms").Cells(i + 1, 6) = breaks(i)
        ThisWorkbook.Sheets("histograms").Cells(i + 1, 7) = freq(i) / intMCIterations 'put into propor
tions
        If i = 1 Then
            ThisWorkbook.Sheets("histograms").Cells(i + 1, 8) = Cells(i + 1, 7)
        Else
            ThisWorkbook.Sheets("histograms").Cells(i + 1, 8) = ThisWorkbook.Sheets("histograms").
Cells(i, 8) + ThisWorkbook.Sheets("histograms").Cells(i + 1, 7) '[cab] cdf for schedule
        End If
    Next i

    '[cab] Generate Interactive Chart

    Dim CostChart As ChartObject
    Set CostChart = ThisWorkbook.Sheets("histograms").ChartObjects.Add(1010, 5, 400, 300) 'left, top,
width, height relative to A1
    With CostChart.chart
        .SetSourceData Source:=Range(ThisWorkbook.Sheets("histograms").Cells(2, 7), ThisWorkbook.Sheet
s("histograms").Cells(intBins + 1, 8)) 'get the two series
        .SeriesCollection(1).XValues = Range(ThisWorkbook.Sheets("histograms").Cells(2, 6), (ThisWorkb
ook.Sheets("histograms").Cells(intBins + 1, 6)))
        .SeriesCollection(2).XValues = Range(ThisWorkbook.Sheets("histograms").Cells(2, 6), (ThisWorkb
ook.Sheets("histograms").Cells(intBins + 1, 6)))
        .chartType = xlXYScatter 'start with scatter chart
        .HasTitle = True
        .HasLegend = False
        .ChartTitle.Caption = "KT Cost Distribution (red=cdf, blue-pdf)"
        .SetElement (msoElementPrimaryValueGridLinesNone) 'turns off gridlines
        With .Axes(xlCategory)
            .HasTitle = True
            .AxisTitle.text = "Dollars (millions)"
            .TickLabelSpacing = 10 'these can be handy for chart scaling in applications
        End With
        With .Axes(xlValue)
            .HasTitle = True
```

```
DataViz - 14
            .AxisTitle.text = "Probability"
            .MinimumScale = 0
            .MaximumScale = 1
            .CrossesAt = 0
            .MajorUnit = 0.2
            .MinorUnit = 0.1
        End With
    End With

    ' [cab] Interpolate to have slider match the curve due to connecting scatter points linearly
    ' Note: Using scatterplots due to x-axis problems using line graphs
    Dim j1 As Integer
    Dim j2 As Integer
    Dim maxj As Integer
    ThisWorkbook.Sheets("histograms").Cells(1, 9) = "costhat"
    ThisWorkbook.Sheets("histograms").Cells(2, 9) = Round(ThisWorkbook.Sheets("histograms").Cells(2, 6
)) - 1
    ThisWorkbook.Sheets("histograms").Cells(1, 10) = "cdfhat"
    ThisWorkbook.Sheets("histograms").Cells(1, 11) = "pdfhat"

    For i = 2 To Round(ThisWorkbook.Sheets("histograms").Cells(intBins + 1, 6))
        If ThisWorkbook.Sheets("histograms").Cells(i, 9) > ThisWorkbook.Sheets("histograms").Cells(int
Bins + 1, 6) Then
            maxj = i
            Exit For
        Else
            ThisWorkbook.Sheets("histograms").Cells(i + 1, 9) = ThisWorkbook.Sheets("histograms").Cell
s(i, 9) + 1
            For j1 = 2 To intBins + 1
                If ThisWorkbook.Sheets("histograms").Cells(j1, 6) > ThisWorkbook.Sheets("histograms").
Cells(i, 9) Then
                    Exit For
                End If
            Next j1
            For j2 = intBins + 1 To 2 Step -1
                If ThisWorkbook.Sheets("histograms").Cells(j2, 6) <= ThisWorkbook.Sheets("histograms")
.Cells(i, 9) Then
                    Exit For
                End If
            Next j2
            If (j1 <> 1 And j2 <> 1) Then
                ThisWorkbook.Sheets("histograms").Cells(i, 10) = ThisWorkbook.Sheets("histograms").Cel
ls(j2, 8) + (ThisWorkbook.Sheets("histograms").Cells(j1, 8) - ThisWorkbook.Sheets("histograms").Cells(
j2, 8)) * ((ThisWorkbook.Sheets("histograms").Cells(i, 9) - ThisWorkbook.Sheets("histograms").Cells(j2
, 6)) / (ThisWorkbook.Sheets("histograms").Cells(j1, 6) - ThisWorkbook.Sheets("histograms").Cells(j2,
6)))
            ElseIf (j1 = intBins + 1 And j2 = intBins + 1) Then
                ThisWorkbook.Sheets("histograms").Cells(i, 10) = ThisWorkbook.Sheets("histograms").Cel
ls(intBins + 1, 8)
            Else
                ThisWorkbook.Sheets("histograms").Cells(i, 10) = ThisWorkbook.Sheets("histograms").Cel
ls(2, 8)
            End If
        End If
    Next i

    ThisWorkbook.Sheets("histograms").Cells(2, 11) = ThisWorkbook.Sheets("histograms").Cells(2, 10)
    For i = 3 To Round(ThisWorkbook.Sheets("histograms").Cells(intBins + 1, 6))
        If ThisWorkbook.Sheets("histograms").Cells(i, 9) > ThisWorkbook.Sheets("histograms").Cells(int
Bins + 1, 6) Then
            Exit For
        Else
            ThisWorkbook.Sheets("histograms").Cells(i, 11) = ThisWorkbook.Sheets("histograms").Cells(i
, 10) - ThisWorkbook.Sheets("histograms").Cells(i - 1, 10)
            If ThisWorkbook.Sheets("histograms").Cells(i, 11) < 0 Then
                ThisWorkbook.Sheets("histograms").Cells(i, 11) = 0
            End If
        End If
    Next i

    With CostChart.chart '.pdf in accordance with displayed axis
```

```
DataViz - 15
        .SeriesCollection(1).Values = Range(ThisWorkbook.Sheets("histograms").Cells(2, 11), ThisWorkbo
ok.Sheets("histograms").Cells(maxj, 11))
        .SeriesCollection(1).XValues = Range(ThisWorkbook.Sheets("histograms").Cells(2, 9), ThisWorkbo
ok.Sheets("histograms").Cells(maxj, 9))
        .SeriesCollection(2).Values = Range(ThisWorkbook.Sheets("histograms").Cells(2, 10), ThisWorkbo
ok.Sheets("histograms").Cells(maxj, 10))
        .SeriesCollection(2).XValues = Range(ThisWorkbook.Sheets("histograms").Cells(2, 9), ThisWorkbo
ok.Sheets("histograms").Cells(maxj, 9))
    End With

'[cab] Put data readout beside the chart
    ThisWorkbook.Sheets("histograms").Cells(22, 24) = "X Value"
    ThisWorkbook.Sheets("histograms").Cells(22, 25) = "Cum. Prob."
    ThisWorkbook.Sheets("histograms").Cells(22, 26) = "Frequency"
    ThisWorkbook.Sheets("histograms").Cells(23, 24) = Cells(2, 6)
    Application.ThisWorkbook.Sheets("histograms").Range("y23").Formula = "=Lookup(x23,i2:i10000,j2:j10
00)" 'actually write formula for dynamic updating
    Application.ThisWorkbook.Sheets("histograms").Range("z23").Formula = "=Lookup(x23,i2:i1000,k2:k100
0)"

    Dim CostScroll As ScrollBar ' NOTE microsoft library has all properties
    Set CostScroll = ThisWorkbook.Sheets("histograms").ScrollBars.Add(1010, 305, 400, 10) 'adds the de
clared scroll bar at those points in sheet
    With CostScroll 'scroll bar properties
        .Value = ThisWorkbook.Sheets("histograms").Cells(2, 6)
        .min = Round(ThisWorkbook.Sheets("histograms").Cells(2, 6)) - 1
        .max = ThisWorkbook.Sheets("histograms").Cells(intBins + 1, 6)
        .SmallChange = 1
        .LargeChange = 10
        .LinkedCell = "=histograms!$x$23"
        .Display3DShading = True
    End With

    With CostChart.chart
        .SeriesCollection.NewSeries 'declares a new series
        With .SeriesCollection(3)
            .chartType = xlXYScatter 'makes it a scatterplot
            .XValues = "=histograms!$x$23" 'x value, linked to scroll bar
            .Values = "=histograms!$y$23" 'y value'
            .MarkerStyle = -4168
            .Format.Fill.Visible = msoFalse
            .Format.Line.Visible = msoTrue
            .Format.Line.ForeColor.ObjectThemeColor = msoThemeColorText1
        End With

    With ThisWorkbook.Sheets("histograms").Range("y23:z23")
        .Style = "Percent"
        .NumberFormat = "0.0%"
    End With

        With .Axes(xlCategory)
            .MinimumScale = Round(ThisWorkbook.Sheets("histograms").Cells(2, 6)) - 1
            .MaximumScale = Round(ThisWorkbook.Sheets("histograms").Cells(intBins + 1, 6))
        End With

        With .SeriesCollection(1)
            .Format.Line.Visible = msoTrue 'formats the pdf series
            .Format.Line.ForeColor.ObjectThemeColor = msoThemeColorAccent1
            .Format.Line.DashStyle = msoLineSysDot
            .MarkerStyle = -4142
        End With

        With .SeriesCollection(2)
            .Format.Line.Visible = msoTrue 'formats the pdf series
            .MarkerStyle = -4142
            .Format.Line.ForeColor.ObjectThemeColor = msoThemeColorAccent2
        End With

    End With

'[cab] Put Scedule Duration Time on Table worksheet
```

```
DataViz - 16

    ThisWorkbook.Sheets("tables").Cells(14, 7) = "Schedule Duration Time"
    ThisWorkbook.Sheets("tables").Cells(15, 7) = dblReqTime
    With ThisWorkbook.Sheets("tables").Range("g14:g15")
        .HorizontalAlignment = xlCenter
    End With

    Application.Calculation = xlAutomatic
End Sub
```

DistributionFunctions Module

This module provides functions used to manipulate the probability density functions and cumulative distribution functions used in RTRAM.

```
DistributionFunctions - 1

Option Explicit

' compute the PDF of a triangle distribution
Function pdfTriangle(X As Double, a As Double, b As Double, c As Double)
    If (X < a) Then
        pdfTriangle = 0
    ElseIf (X <= c And X >= a) Then
        pdfTriangle = (2 * (X - a)) / ((b - a) * (c - a))
    ElseIf (X <= b And X > c) Then
        pdfTriangle = (2 * (b - X)) / ((b - a) * (b - c))
    Else
        pdfTriangle = 0
    End If
End Function

' compute the cdf of a triangle distribution
Function cdfTriangle(X As Double, a As Double, b As Double, c As Double)
    If (X < a) Then
        cdfTriangle = 0
    ElseIf (X <= c And X >= a) Then
        cdfTriangle = (X - a) ^ 2 / ((b - a) * (c - a))
    ElseIf (X <= b And X > c) Then
        cdfTriangle = 1 - (b - X) ^ 2 / ((b - a) * (b - c))
    Else
        cdfTriangle = 1
    End If
End Function

' compute the inverse CDF of a triangle distribution
Function invCDFTriangle(p As Double, a As Double, b As Double, c As Double)
    If (0 = b - a) Then
        invCDFTriangle = 0
    Else
        If (p < (c - a) / (b - a)) Then
            invCDFTriangle = Sqr(p * (c - a) * (b - a)) + a
        Else
            invCDFTriangle = b - Sqr((1 - p) * (b - c) * (b - a))
        End If
    End If
End Function

' calculate an empirical CDF based on a set of samples
Function empiricalCDF(arrSamples() As Variant, t As Variant)
    Dim intCount As Integer
    intCount = 0

    Dim dblSample As Variant
    For Each dblSample In arrSamples
        If (dblSample <= t) Then
            intCount = intCount + 1
        End If
    Next dblSample

    empiricalCDF = intCount / (UBound(arrSamples) - LBound(arrSamples) + 1)
End Function
```

Helpers Module

This module contains a handful of utility functions that are useful across modules. For example, one function will perform quicksort on a VBA array.

```
Helpers - 1

Option Explicit

' Here are some worksheet variables
Public wsInputUI1 As Worksheet
Public wsAlternatives As Worksheet
Public wsTechnologies As Worksheet
Public wsDurations As Worksheet
Public wsOutputDistributions As Worksheet
Public wsOutputSchedule As Worksheet
Public wsOutputConsequence As Worksheet
Public wsOutputUI1 As Worksheet
Public wsOutput3d As Worksheet

Sub setSheets()
    Set wsInputUI1 = Sheets("UI-1")
    Set wsAlternatives = Sheets("Alternatives")
    Set wsTechnologies = Sheets("Key Technologies")
    Set wsDurations = Sheets("Schedule Durations")
    Set wsOutputDistributions = Sheets("Distributions")
    Set wsOutputSchedule = Sheets("Schedule")
    Set wsOutputConsequence = Sheets("Model Output")
    Set wsOutputUI1 = Sheets("Histogram")
    Set wsOutput3d = Sheets("Bubble Chart")
End Sub

' right pad text string with specified character
Function PadRight(text As Variant, lengthLimit As Integer, padCharacter As String) As String
    If (Len(text) < lengthLimit) Then
        PadRight = CStr(text) & String(lengthLimit - Len(CStr(text)), padCharacter)
    Else
        PadRight = Left(CStr(text), lengthLimit)
    End If
End Function

' implementation of quicksort algorithm
Sub QuickSort(vArray As Variant, inLow As Long, inHi As Long)

    Dim pivot    As Variant
    Dim tmpSwap As Variant
    Dim tmpLow  As Long
    Dim tmpHi    As Long

    tmpLow = inLow
    tmpHi = inHi

    pivot = vArray((inLow + inHi) \ 2)

    While (tmpLow <= tmpHi)

        While (vArray(tmpLow) < pivot And tmpLow < inHi)
            tmpLow = tmpLow + 1
        Wend

        While (pivot < vArray(tmpHi) And tmpHi > inLow)
            tmpHi = tmpHi - 1
        Wend

        If (tmpLow <= tmpHi) Then
            tmpSwap = vArray(tmpLow)
            vArray(tmpLow) = vArray(tmpHi)
            vArray(tmpHi) = tmpSwap
            tmpLow = tmpLow + 1
            tmpHi = tmpHi - 1
        End If

    Wend

    If (inLow < tmpHi) Then QuickSort vArray, inLow, tmpHi
    If (tmpLow < inHi) Then QuickSort vArray, tmpLow, inHi

End Sub
```

```
Helpers - 2

Function collectionToArray(c As Collection) As Variant()
    Dim a() As Variant: ReDim a(0 To c.Count - 1)
    Dim i As Integer
    For i = 1 To c.Count
        a(i - 1) = c.Item(i)
    Next
    collectionToArray = a
End Function

Sub ExportWorksheet()
    Dim ws As Worksheet
    Set ws = ActiveSheet

    Dim wbkcopy As Workbook 'dummy workbook to keep original name
    Set wbkcopy = Workbooks.Add(1) 'One sheet dummy workbook

    ws.Copy Before:=wbkcopy.Sheets(1) 'Note the tab name being copied

    Dim Filename As Variant
    Filename = Application.GetSaveAsFilename(, "Comma Separated Value File (*.csv), *.csv") 'GetSaveAs
Filename triggers browsing menu
    If Filename = False Then 'If no name entered
        Application.DisplayAlerts = False 'Turns off annoying "Are you sure you want to close"?
        wbkcopy.Close
        Exit Sub
        Application.DisplayAlerts = True
    End If

    wbkcopy.SaveAs Filename, xlCSV
    Application.DisplayAlerts = False 'Turns off annoying "Are you sure you want to close"?
    wbkcopy.Close
    Application.DisplayAlerts = True
End Sub
```

RTRAM Module

The RTRAM module contains the code that performs the MC simulation; calculates performance, schedule, and cost outcomes; and calls the routines that write results to the model output worksheets. The code is as follows:

```
RTRAM - 1

Option Explicit

' Here are some model variables
Private intSimDurationInMonths As Integer
Private intNumMCIterations As Integer
Private dblTolerance As Double

Sub clearSheets()
    wsOutputDistributions.Cells.Clear
    With wsOutputDistributions
        .Cells(1, 1) = "Alternative"
        .Cells(1, 2) = "Key Technology"
        .Cells(1, 3) = "COA"
        .Cells(1, 4) = "Schedule Duration"
        .Cells(1, 5) = "Month"
        .Cells(1, 6) = "g hat"
        .Cells(1, 7) = "G"
    End With

    wsOutputSchedule.Cells.Clear
    With wsOutputSchedule
        .Cells(1, 1) = "Alternative"
        .Cells(1, 2) = "Technology"
        .Cells(1, 3) = "COA"
        .Cells(1, 4) = "Schedule Duration"
        .Cells(1, 5) = "mcIteration"
        .Cells(1, 6) = "Delivered"
    End With

    wsOutputConsequence.Cells.Clear
    With wsOutputConsequence
        .Cells(1, 1) = "Alternative"
        .Cells(1, 2) = "Schedule Duration"
        .Cells(1, 3) = "mcIteration"
        .Cells(1, 4) = "cPerf"
        .Cells(1, 5) = "cSched"
        .Cells(1, 6) = "cCost"
    End With
End Sub

Sub writeOutputDistribution(Alt As Alternative, KeyTech As KeyTechnology, SchedDuration As Double, Mon
th As Integer, gHat As Double, g As Double)
    With wsOutputDistributions
        Dim RowNum As Long
        RowNum = .Cells(Rows.Count, "A").End(xlUp).row
        .Cells(RowNum + 1, 1) = Alt.Name
        .Cells(RowNum + 1, 2) = KeyTech.Name
        .Cells(RowNum + 1, 3) = Alt.COAs(KeyTech.Name)
        .Cells(RowNum + 1, 4) = SchedDuration
        .Cells(RowNum + 1, 5) = Month
        .Cells(RowNum + 1, 6) = gHat
        .Cells(RowNum + 1, 7) = g
    End With
End Sub

Sub writeOutputSchedule(Alt As Alternative, KeyTech As KeyTechnology, SchedDuration As Double, mcItera
tion As Integer, Delivered As Double)
    With wsOutputSchedule
        Dim RowNum As Long
        RowNum = .Cells(Rows.Count, "A").End(xlUp).row
        .Cells(RowNum + 1, 1) = Alt.Name
        .Cells(RowNum + 1, 2) = KeyTech.Name
        .Cells(RowNum + 1, 3) = Alt.COAs(KeyTech.Name)
        .Cells(RowNum + 1, 4) = SchedDuration
        .Cells(RowNum + 1, 5) = mcIteration
        .Cells(RowNum + 1, 6) = Delivered
    End With
End Sub

Sub writeOutputConsequence(Alt As Alternative, SchedDuration As Double, mcIteration As Integer, cPerf
As Double, cSched As Double, cCost As Double)
```

```
RTRAM - 2

    With wsOutputConsequence
        Dim RowNum As Long
        RowNum = .Cells(Rows.Count, "A").End(xlUp).row
        .Cells(RowNum + 1, 1) = Alt.Name
        .Cells(RowNum + 1, 2) = SchedDuration
        .Cells(RowNum + 1, 3) = mcIteration
        .Cells(RowNum + 1, 4) = cPerf
        .Cells(RowNum + 1, 5) = cSched
        .Cells(RowNum + 1, 6) = cCost
    End With
End Sub

' compute little G - see RTRAM model write-up
Function computeLittleG(KeyTech As KeyTechnology, intSimDurationInMonths As Integer, intNumMCIteration
s As Integer)
    Dim arrTRL() As Double
    Dim arrMRLCond() As Double
    Dim arrIRLCond() As Double
    Dim arrGLittle() As Variant

    ReDim arrTRL(1 To intNumMCIterations)
    ReDim arrMRLCond(1 To intNumMCIterations)
    ReDim arrIRLCond(1 To intNumMCIterations)
    ReDim arrGLittle(1 To intNumMCIterations)

    ' COMPUTE: g_nk(t), the tech specific sched implied by risk wkshop

    Dim i As Integer, p As Double
    For i = 1 To intNumMCIterations
        ' compute t_TRL
        p = Rnd()
        arrTRL(i) = invCDFTriangle(p, KeyTech.TRLmin, KeyTech.TRLmax, KeyTech.TRLmode)

        ' compute t_MRL|TRL
        p = Rnd()
        If (KeyTech.MRLmin = KeyTech.MRLmax) Then
            arrMRLCond(i) = KeyTech.MRLmin
        Else
            arrMRLCond(i) = invCDFTriangle(p, KeyTech.MRLmin, KeyTech.MRLmax, KeyTech.MRLmode)
        End If

        ' compute t_IRL|TRL
        p = Rnd()
        If (KeyTech.IRLmin = KeyTech.IRLmax) Then
            arrIRLCond(i) = KeyTech.IRLmin
        Else
            arrIRLCond(i) = invCDFTriangle(p, KeyTech.IRLmin, KeyTech.IRLmax, KeyTech.IRLmode)
        End If

        ' compute max(t_MRL|TRL, t_IRL|TRL)
        arrGLittle(i) = Application.WorksheetFunction.max(arrTRL(i) + arrMRLCond(i), arrTRL(i) + arrIR
LCond(i))
    Next i

    ' COMPUTE: g "hat"

    Dim arrgHat() As Double
    ReDim arrgHat(1 To intSimDurationInMonths)

    For i = 1 To intSimDurationInMonths
        arrgHat(i) = empiricalCDF(arrGLittle, CDbl(i)) - empiricalCDF(arrGLittle, CDbl(i - 1))
    Next i

    computeLittleG = arrgHat
End Function

Sub GetModelRuns()
    Application.ScreenUpdating = False
    Application.Calculation = xlManual

    Call setSheets
```

```
RTRAM - 3

    Call clearSheets

    ' these are some handy index variables
    Dim i As Integer, j As Integer, k As Integer
    Dim oleObj As Object

    ' model parameters
    '---------------------------------------------------------------------------

    ' Number of Monte Carlo iterations per alterantive
    With wsInputUI1.OLEObjects("Num Iterations value").Object
        If .Value = vbNullString Then
            MsgBox "Num Iterations must be a number"
            End
        End If

        If Not IsNumeric(.Value) And .Value <> vbNullString Then
            MsgBox "Num Iterations must be a number"
            End
        End If
        intNumMCIterations = CInt(.Value)
    End With

    With wsInputUI1.OLEObjects("Tolerance value").Object
        If .Value = vbNullString Then
            MsgBox "Tolerance must be a number"
            End
        End If

        If Not IsNumeric(.Value) And .Value <> vbNullString Then
            MsgBox "Tolerance must be a number"
            End
        End If
        dblTolerance = CDbl(.Value)
    End With

    ' need to make sure alts and techs don't get out of order
    '---------------------------------------------------------------------------

    Dim NumAlternatives As Integer
    With wsAlternatives
        NumAlternatives = .Cells(Rows.Count, "A").End(xlUp).row - 1
    End With

    Dim AllAlternatives() As String
    For i = 1 To NumAlternatives
        ReDim Preserve AllAlternatives(1 To i)
        AllAlternatives(i) = wsAlternatives.Cells(i + 1, "A").Value
    Next i

    Dim NumTechnologies As Integer
    With wsTechnologies
        NumTechnologies = .Cells(Rows.Count, "A").End(xlUp).row - 1
    End With

    Dim AllTechnologies() As String
    For i = 1 To NumTechnologies
        ReDim Preserve AllTechnologies(1 To i)
        AllTechnologies(i) = wsTechnologies.Cells(i + 1, "A").Value
    Next i

    Dim NumDurations As Integer
    With wsDurations
        NumDurations = .Cells(Rows.Count, "A").End(xlUp).row - 1
    End With

    Dim AllDurations() As Double
    For i = 1 To NumDurations
        ReDim Preserve AllDurations(1 To i)
        AllDurations(i) = wsDurations.Cells(i + 1, "A").Value
    Next i
```

RTRAM - 4

```
' put key technologies into a collection
'------------------------------------------------------------------------
Dim Technology As KeyTechnology
Dim Technologies As New Collection
For i = 1 To NumTechnologies
    Set Technology = New KeyTechnology
    With Technology
        .Name = wsTechnologies.Cells(i + 1, "A")
        .TRLmin = wsTechnologies.Cells(i + 1, "B")
        .TRLmax = wsTechnologies.Cells(i + 1, "C")
        .TRLmode = wsTechnologies.Cells(i + 1, "D")
        .MRLmin = wsTechnologies.Cells(i + 1, "E")
        .MRLmax = wsTechnologies.Cells(i + 1, "F")
        .MRLmode = wsTechnologies.Cells(i + 1, "G")
        .IRLmin = wsTechnologies.Cells(i + 1, "H")
        .IRLmax = wsTechnologies.Cells(i + 1, "I")
        .IRLmode = wsTechnologies.Cells(i + 1, "J")
        .PerfConsequence = wsTechnologies.Cells(i + 1, "K")
        .ShedConsequence = 0
        .CostConsequence = 0
        .CPlusFixed = wsTechnologies.Cells(i + 1, "L")
        .CPlusVariable = wsTechnologies.Cells(i + 1, "M")
        .CMinusFixed = wsTechnologies.Cells(i + 1, "N")
        .CMinusVariable = wsTechnologies.Cells(i + 1, "O")
    End With
    Technologies.Add Technology, CStr(i)

    ' compute sim duration: max of latest possible delivery date of all KTs
    If (Technology.TRLmax + Technology.MRLmax + Technology.IRLmax > intSimDurationInMonths) Then
        intSimDurationInMonths = Technology.TRLmax + Technology.MRLmax + Technology.IRLmax
    End If
Next i

' compute sim duration: round up to integer number of months
intSimDurationInMonths = Application.RoundUp(intSimDurationInMonths, 0)

' compute g hat for each key technology
For Each Technology In Technologies
    Dim arrgHat() As Double
    arrgHat = computeLittleG(Technology, intSimDurationInMonths, intNumMCIterations)
    Technology.gHat = arrgHat
Next Technology

' get selected alternatives
'------------------------------------------------------------------------
Dim SelectedAlternatives As New Collection
For i = 1 To NumAlternatives
    If (wsInputUI1.OLEObjects("A: " & i).Object.Value) Then
        SelectedAlternatives.Add (i)
    End If
Next i

' ERROR checking
If (SelectedAlternatives.Count = 0) Then
    MsgBox "You must select an alternative"
    End
End If

' get key technologies for selected alternatives
'------------------------------------------------------------------------
Dim Alt As Alternative
Dim Alternatives As New Collection
For i = 1 To SelectedAlternatives.Count

    ' create a new alternative
    Set Alt = New Alternative
    With Alt
        .Name = AllAlternatives(SelectedAlternatives.Item(i))
        Set .KeyTechnologies = New Collection
        Set .COAs = New Collection
```

```
RTRAM - 5

      End With

      ' we need the index number for the alternative
      Dim AltIdx As Integer: AltIdx = SelectedAlternatives.Item(i)

      ' get the KTs and COAs
      For j = 1 To NumTechnologies
          If (wsInputUI1.OLEObjects("AT: " & AltIdx & ", " & j).Object.Value) Then
              Set Technology = Technologies.Item(CStr(j))
              Alt.KeyTechnologies.Add Technology

              ' add COA to the alternative
              Dim COA As String
              COA = wsInputUI1.OLEObjects("AT COA: " & AltIdx & ", " & j).Object.Value
              If (COA = "") Then
                  MsgBox "Must select COA for " & Alt.Name & ", " & Technology.Name
                  End
              End If
              Alt.COAs.Add COA, key:=Technology.Name
          End If
      Next j

      ' ERROR checking
      If (Alt.KeyTechnologies.Count = 0) Then
          MsgBox "You must select technologies for " & Alt.Name
          End
      End If

      Alternatives.Add Alt, key:=Alt.Name
  Next i

  ' get durations
  '--------------------------------------------------------------------------
  Dim SelectedDurations As New Collection
  For i = 1 To NumDurations
      If (wsInputUI1.OLEObjects("D: " & i).Object.Value) Then
          SelectedDurations.Add AllDurations(i)
      End If
  Next i

  ' ERROR checking
  If (SelectedDurations.Count = 0) Then
      MsgBox "You must select a duration"
      End
  End If

  ' run each alternative
  '--------------------------------------------------------------------------
  Dim Duration As Variant
  For Each Duration In SelectedDurations
      For Each Alt In Alternatives
          Call RunAlternative(Alt, CDbl(Duration))
      Next Alt
  Next Duration

  Application.Calculation = xlAutomatic
End Sub

' run the RTRAM model for a single alternative
Sub RunAlternative(Alt As Alternative, SchedDuration As Double)
  Application.ScreenUpdating = False
  Application.Calculation = xlManual

  Dim boolVerbose As Boolean
  boolVerbose = True

  ' these are some handy index variables
  Dim i As Integer, j As Integer, k As Integer
  Dim KeyTech As KeyTechnology

  ' COMPUTE: g, pi, and f
```

RTRAM - 6

```
'--------------------------------------------------------------------------

Dim arrTStar() As Double
ReDim arrTStar(1 To Alt.KeyTechnologies.Count)

'[cab] cumulative distribution from ghat
Dim matGcdf() As Double
ReDim matGcdf(1 To intSimDurationInMonths, 1 To Alt.KeyTechnologies.Count)

i = 0
For Each KeyTech In Alt.KeyTechnologies
    i = i + 1

    ' RECALL: g_nk(t)
    Dim arrgHat() As Double
    arrgHat = KeyTech.gHat

    ' COMPUTE: t*, the maximum T for a particular KT
    Dim dblGHat As Double
    dblGHat = 0
    For j = 1 To UBound(arrgHat)
        dblGHat = dblGHat + arrgHat(j)
        If (Abs(dblGHat - 1) < dblTolerance) Then Exit For
    Next j

    arrTStar(i) = j

    ' COMPUTE: The cdf from g_nk(t)
    matGcdf(1, i) = arrgHat(1)
    For j = 2 To intSimDurationInMonths
        If (j <= UBound(arrgHat)) Then
            matGcdf(j, i) = matGcdf(j - 1, i) + arrgHat(j)
        Else
            matGcdf(j, i) = matGcdf(j - 1, i)
        End If
    Next j

    ' Writes the distributions to intermediate tab - change to include g() and G() only
    If (boolVerbose) Then
        For j = 1 To intSimDurationInMonths
            Call writeOutputDistribution(Alt, KeyTech, SchedDuration, j, arrgHat(j), matGcdf(j, i)
)
        Next j
    End If
Next KeyTech

' SIMULATE the alternative
'--------------------------------------------------------------------------

Dim p As Double
Dim arrOpTime() As Integer
ReDim arrOpTime(1 To Alt.KeyTechnologies.Count)
Dim matRealized() As Double
ReDim matRealized(1 To intNumMCIterations, 1 To Alt.KeyTechnologies.Count)
Dim matTimeDraw() As Double
ReDim matTimeDraw(1 To intNumMCIterations, 1 To Alt.KeyTechnologies.Count)
Dim arrOpReqTime() As Integer
ReDim arrOpReqTime(1 To intNumMCIterations)

For i = 1 To intNumMCIterations

    '[cab] I realized we didn't need to sample from F+ and F- since they
    ' were just conditionals built out of G(). Doing so saves us building
    ' multiple F()s when we have COAEs, and actually is a bit quicker. So
    ' instead, we draw from G(), adjust required time if nec, and form the
    ' realized matrix. Actual assignmnet for histograms is done in
    ' consequence section.

    '[cab] Readiness time is drawn from appropriate G_nk
    '[cab] Here, we build the time draw matrix and redefine operational required time if nec (i.e.
, for COAEs)
```

```
RTRAM - 7

        j = 0
        For Each KeyTech In Alt.KeyTechnologies
            j = j + 1
            matRealized(i, j) = 0 'initialize
            p = Rnd()
            k = 1
            While (matGcdf(k, j) <= p) ' sample from G()
                k = k + 1
            Wend
            matTimeDraw(i, j) = k

            ' Use time draw matrix to determine Operational Required Time if COAE's exist
            arrOpTime(j) = SchedDuration 'Default is the original schedule duration
            If ("COA E" = Alt.COAs(KeyTech.Name) And matTimeDraw(i, j) > SchedDuration) Then 'If need
more time on COAE kt
                arrOpTime(j) = matTimeDraw(i, j) 'Then time draw tells us how much
            End If
        Next KeyTech

        arrOpReqTime(i) = Application.WorksheetFunction.max(arrOpTime) 'Operational required time is m
ax of COAE draws (=dblReqTime if no COAEs)

        '[cab] Build the Realization matrix
        For j = 1 To Alt.KeyTechnologies.Count
            If (matTimeDraw(i, j) <= arrOpReqTime(i)) Then
                matRealized(i, j) = 1 '=1 if "on time" relative to operational time, 0 otherwise
            End If
        Next j
    Next i

    ' COMPUTE: performance consequence
    '-------------------------------------------------------------------------

    Dim dblPerf() As Double
    ReDim dblPerf(1 To intNumMCIterations)
    For i = 1 To intNumMCIterations
        Dim dblPerfRealized() As Double
        ReDim dblPerfRealized(1 To Alt.KeyTechnologies.Count)
        j = 0
        For Each KeyTech In Alt.KeyTechnologies
            j = j + 1
            If (1 = matRealized(i, j)) Then
                dblPerfRealized(j) = 0
            ElseIf ("COA R" = Alt.COAs(KeyTech.Name)) Then '[cab] if not delivered and COA=R then assi
gn consequence
                dblPerfRealized(j) = KeyTech.PerfConsequence
            Else '[cab] if not delivered and not COA=R then consequence=0
                dblPerfRealized(j) = 0
            End If
        Next KeyTech
        dblPerf(i) = Application.WorksheetFunction.max(dblPerfRealized)
    Next i

    ' COMPUTE: schedule and cost consequence
    '-------------------------------------------------------------------------

    Dim dblSched() As Double
    ReDim dblSched(1 To intNumMCIterations)
    Dim dblCost() As Double
    ReDim dblCost(1 To intNumMCIterations)
    For i = 1 To intNumMCIterations
        Dim dblSchedRealized() As Double
        ReDim dblSchedRealized(1 To Alt.KeyTechnologies.Count)
        Dim dblCostRealized() As Double
        ReDim dblCostRealized(1 To Alt.KeyTechnologies.Count)

        Dim dblT As Double '[cab] moved out of loop
        Dim dblTMinus As Double '[cab] moved out of loop
        j = 0
        For Each KeyTech In Alt.KeyTechnologies
            j = j + 1
```

```
RTRAM - 8

            If (1 = matRealized(i, j)) Then 'kt j is delivered. Incld. random draws and all COAE techn
ologies
                dblT = matTimeDraw(i, j)
                Call writeOutputSchedule(Alt, KeyTech, SchedDuration, i, dblT)

                dblSchedRealized(j) = dblT
                dblCostRealized(j) = KeyTech.CPlusFixed + dblT * KeyTech.CPlusVariable

            ElseIf ("COA F" = Alt.COAs(KeyTech.Name)) Then   ' [cab] kt j not delivered and COAF
                'Debug.Print "Reached a COA F cost calculation"
                dblT = matTimeDraw(i, j)
                Call writeOutputSchedule(Alt, KeyTech, SchedDuration, i, dblT)

                dblCostRealized(j) = KeyTech.CPlusFixed + dblT * KeyTech.CPlusVariable   '[cab] cost eq
. to draw from non-delivered distribution...

                If (1 = KeyTech.ShedConsequence) Then
                    dblTMinus = arrTStar(j)
                Else
                    dblTMinus = arrOpReqTime(i)   'Sched=0 on input tab suggests use OpReqTime
                End If

                dblSchedRealized(j) = dblTMinus '[cab] ...but delivered at point in time as specified
by user choice on inputs

            Else 'kt not delivered and COAR
                'Debug.Print "Reached a COA R cost calculation"
                If (1 = KeyTech.ShedConsequence) Then
                    dblTMinus = arrTStar(j)
                Else
                    dblTMinus = arrOpReqTime(i)
                End If
                Call writeOutputSchedule(Alt, KeyTech, SchedDuration, i, dblTMinus)

                dblSchedRealized(j) = dblTMinus
                dblCostRealized(j) = KeyTech.CPlusFixed + arrOpReqTime(i) * KeyTech.CPlusVariable + Ke
yTech.CMinusFixed + (dblTMinus - arrOpReqTime(i)) * KeyTech.CMinusVariable
            End If
        Next KeyTech

        dblSched(i) = Application.WorksheetFunction.max(dblSchedRealized)
        For j = 1 To Alt.KeyTechnologies.Count
            dblCost(i) = dblCost(i) + dblCostRealized(j)
        Next j
    Next i

    ' write out intermediate results to worksheets
    '-------------------------------------------------------------------------

    If (boolVerbose) Then
        For i = 1 To intNumMCIterations
            Call writeOutputConsequence(Alt, SchedDuration, i, dblPerf(i), dblSched(i), dblCost(i))
        Next i
    End If

    Application.Calculation = xlAutomatic
End Sub
```

Class Modules

RTRAM defines four class modules that define object properties and methods used in other parts of the code.

Alternative Class Module

```
Alternative - 1

Option Explicit

Public Name As String
Public KeyTechnologies As Collection
Public COAs As Collection
```

ChartSeries Module

```
ChartSeries - 1

Option Explicit

Public seriesName As String
Public SeriesData As Collection
```

Key Technology Class Module

```
KeyTechnology - 1

Option Explicit

Public Name As String

' TRL triangular distribution parameters
Public TRLmin As Double
Public TRLmax As Double
Public TRLmode As Double

' MRL triangular distribution parameters
Public MRLmin As Double
Public MRLmax As Double
Public MRLmode As Double

' IRL triangular distribution parameters
Public IRLmin As Double
Public IRLmax As Double
Public IRLmode As Double

' Consequences
Public PerfConsequence As Double
Public ShedConsequence As Double
Public CostConsequence As Double

' Cost parameters
Public CPlusFixed As Double
Public CPlusVariable As Double
Public CMinusFixed As Double
Public CMinusVariable As Double

' g hat
Private g() As Double

Public Property Get gHat() As Double()
    gHat = g
End Property

Public Property Let gHat(gHat() As Double)
    g = gHat
End Property
```

RTRAMResults Class Module

```
RTRAMResults - 1

Option Explicit

Public ScheduleDuration As Double
Public cPerf As Variant
Public cSchd As Variant
Public cCost As Variant
```

Variable Names in the Risk-Informed Trade Analysis Model

Table D.1 provides a bridge between the notation used in Chapter Five and this appendix and the computer code.

Table D.1
Variable Names Used in the Risk-Informed Trade Analysis Model Code

Description	Documentation Variable	Code Variable	Notes
KT-specific schedule pdf	$g_{nk}(t)$	arrgHat	
KT-specific schedule cdf	Not applicable	matGcdf	
KT-specific schedule draw	t_{nk}	matTimeDraw	Draw from $g_{nk}(t)$
Milestone date	ms	SchedDuration	
Maximum allowable KT-specific schedule	Not applicable	arrOpTime	= milestone date if $coa \neq coa^e$; t_{nk} otherwise
Implied maximum system schedule time	$\max(arrOpTime)$	arrOpReqTime	
KT-specific performance consequence	$p_{nk}\left(kt_{nk}^{i}\right)$	dblPerfRealized	= 0 if KT delivered
System performance outcome	P_n	dblPerf	$= \max(p_{n1},...,p_{nK})$
KT-specific schedule outcome	rt_{nk}	dblSchedRealized	
System schedule outcome	rt_n	dblSched	$= \max(rt_{n1},...,rt_{nK})$
KT-specific cost outcome	c_{nk}^{i}	dblCostRealized	
System cost outcome	c_{nk}	dblCost	$= \sum_k c_{nk}^{i}$

NOTE: pdf = probability distribution function. cdf = cumulative distribution function.

References

Alexander, Arthur J., and J. R. Nelson, *Measuring Technological Change: Aircraft Turbine Engines*, Santa Monica, Calif.: RAND Corporation, R-1017-ARPA/PR, 1972. As of November 5, 2013:
http://www.rand.org/pubs/reports/R1017.html

AMSAA—*See* U.S. Army Materiel Systems Analysis Activity.

Arena, Mark V., Robert S. Leonard, Sheila E. Murray, and Obaid Younossi, *Historical Cost Growth of Completed Weapon System Programs*, Santa Monica, Calif.: RAND Corporation, TR-343-AF, 2006. As of November 5, 2013:
http://www.rand.org/pubs/technical_reports/TR343.html

Armstrong, J. Scott, "Combining Forecasts," in Jon Scott Armstrong, ed., *Principles of Forecasting: A Handbook for Researchers and Practitioners*, Boston, Mass.: Kluwer Academic, 2001, pp. 417–440.

Ayyub, Bilal M., *Elicitation of Expert Opinions for Uncertainty and Risks*, Boca Raton, Fla.: CRC Press, 2001a.

———, *A Practical Guide on Conducting Expert-Opinion Elicitation of Probabilities and Consequences for Corps Facilities*, Alexandria, Va.: U.S. Army Corps of Engineers Institute for Water Resources, IWR Report 01-R-01, January 2001b. As of November 5, 2013:
http://www.iwr.usace.army.mil/Portals/70/docs/iwrreports/01-R-01.pdf

Bolten, Joseph G., Robert S. Leonard, Mark V. Arena, Obaid Younossi, and Jerry M. Sollinger, *Sources of Weapon System Cost Growth: Analysis of 35 Major Defense Acquisition Programs*, Santa Monica, Calif.: RAND Corporation, MG-670-AF, 2008. As of November 5, 2013:
http://www.rand.org/pubs/monographs/MG670.html

Bounds, Thomas, *Army Independent Risk Assessment Guidebook*, U.S. Army Materiel Systems Analysis Activity, TR-2014-19, April 2014.

Brown, Bernice B., *Delphi Process: A Methodology Used for the Elicitation of Opinions of Experts*, Santa Monica, Calif.: RAND Corporation, P-3925, 1968. As of August 27, 2014:
http://www.rand.org/pubs/papers/P3925.html

CAPE—*See* Office of Cost Assessment and Program Evaluation.

CEAC—*See* U.S. Army Cost and Economic Analysis Center.

Clemen, Robert T., "Combining Forecasts: A Review and Annotated Bibliography," *International Journal of Forecasting*, Vol. 5, No. 4, 1989, pp. 559–583.

Clemen, Robert T., and Robert L. Winkler, "Combining Probability Distributions from Experts in Risk Analysis," *Risk Analysis*, Vol. 19, No. 2, April 1999, pp. 187–203.

Covert, Raymond P., *Analytic Method for Probabilistic Cost and Schedule Risk Analysis: Final Report*, Washington, D.C.: National Aeronautics and Space Administration Office of Program Analysis and Evaluation Cost Analysis Division, April 5, 2013. As of September 30, 2013:
http://www.nasa.gov/pdf/741989main_Analytic%20Method%20for%20Risk%20Analysis%20-%20Final%20Report.pdf

Dana, Jason, and Robyn M. Dawes, "The Superiority of Simple Alternatives to Regression for Social Science Predictions," *Journal of Educational and Behavioral Statistics*, Vol. 29, No. 3, Autumn 2004, pp. 317–331.

Daskilewicz, Matthew J., Brian J. German, Timothy T. Takahashi, Shane Donovan, and Arvin Shajanian, "Effects of Disciplinary Uncertainty on Multi-Objective Optimization in Aircraft Conceptual Design," *Structural and Multidisciplinary Optimization*, Vol. 44, No. 6, 2011, pp. 831–846.

Galway, Lionel A., *Subjective Probability Distribution Elicitation in Cost Risk Analysis: A Review*, Santa Monica, Calif.: RAND Corporation, TR-410-AF, 2007. As of November 6, 2113:
http://www.rand.org/pubs/technical_reports/TR410.html

Garvey, Paul R., "A Family of Joint Probability Models for Cost and Schedule Uncertainties," presented at the 27th Annual Department of Defense Cost Analysis Symposium, September 1993. As of August 27, 2014:
http://www.dtic.mil/dtic/tr/fulltext/u2/a275513.pdf

————, *Probability Methods for Cost Uncertainty Analysis: A Systems Engineering Perspective*, New York: CRC Press, 2000.

Hammitt, James K., and Yifan Zhang, "Combining Experts' Judgments: Comparison of Algorithmic Methods Using Synthetic Data," *Risk Analysis*, Vol. 33, No. 1, January 2013, pp. 109–120.

Hastie, Reid, and Robyn M. Dawes, *Rational Choice in an Uncertain World: The Psychology of Judgment and Decision Making*, 2nd ed., SAGE Publications, 2010.

Henry, T. M., "Technical Risk Assessment Methodology," briefing, Aberdeen, Md.: U.S. Army Materiel Systems Analysis Activity, August 16, 2012.

Kahneman, Daniel, Paul Slovic, and Amos Tversky, eds., *Judgment Under Uncertainty: Heuristics and Biases*, Cambridge University Press, 1982.

Lurie, Philip M., and Matthew S. Goldberg, *A Method for Simulating Correlated Random Variables from Partially Specified Distributions*, Alexandria, Va.: Institute for Defense Analysis, IDA Paper P-2998, October 1994. As of August 27, 2014: http://www.dtic.mil/docs/citations/ADA288820

Mas-Colell, Andreu, Michael Dennis Whinston, and Jerry R. Green, *Microeconomic Theory*, New York: Oxford University Press, 1995.

Mattson, Christopher A., and Achille Messac, "Pareto Frontier Based Concept Selection Under Uncertainty, with Visualization," *Optimization and Engineering*, Vol. 6, No. 1, 2005, pp. 85–115.

Morgan, Millett Granger, and Max Henrion, *Uncertainty: A Guide to Dealing with Uncertainty in Quantitative Risk and Policy Analysis*, Cambridge, UK: Cambridge University Press, 1990, pp. 102–171.

Nelson, J. R., *Performance/Schedule/Cost Tradeoffs and Risk Analysis for the Acquisition of Aircraft Turbine Engines: Applications of R-1288-PR Methodology*, Santa Monica, Calif.: RAND Corporation, R-1781-PR, 1975. As of August 27, 2014: http://www.rand.org/pubs/reports/R1781.html

Nelson, J. R., and Fred Timson, *Relating Technology to Acquisition Costs: Aircraft Turbine Engines*, Santa Monica, Calif.: RAND Corporation, R-1288-PR, 1974. As of August 27, 2014: http://www.rand.org/pubs/reports/R1288.html

"Nunn-McCurdy (NM) Unit Cost Breaches," May 2002. As of August 28, 2014: http://www.defense.gov/news/May2002/d20020502nmc.pdf

Office of Cost Assessment and Program Evaluation, "Analysis of Alternative Studies—Discussion with Key Stakeholders," brief delivered to key analysis-of-alternatives stakeholders, Washington, D.C., February 1, 2013.

Office of the Under Secretary of Defense for Acquisition, Technology, and Logistics, *Performance of the Defense Acquisition System: 2013 Annual Report*, June 28, 2013. As of September 19, 2013: http://www.acq.osd.mil/docs/Performance%20of%20the%20Def%20Acq%20System%202013%20-%20FINAL%2028June2013.pdf

OUSD(AT&L)—*See* Office of the Under Secretary of Defense for Acquisition, Technology, and Logistics.

Public Law 97-86, National Defense Authorization Act for Fiscal Year 1982, December 1, 1981. As of August 28, 2014: http://www.gpo.gov/fdsys/pkg/STATUTE-95/pdf/STATUTE-95-Pg1099.pdf

Public Law 111-23, Weapon Systems Acquisition Reform Act of 2009, May 22, 2009. As of August 28, 2014:
http://www.gpo.gov/fdsys/pkg/PLAW-111publ23/content-detail.html

Sackman, Harold, *Delphi Assessment: Expert Opinion, Forecasting, and Group Process*, Santa Monica, Calif.: RAND Corporation, R-1283-PR, 1974. As of August 27, 2014:
http://www.rand.org/pubs/reports/R1283.html

Shishko, Robert, *Technological Change Through Product Improvement in Aircraft Turbine Engines*, Santa Monica, Calif.: RAND Corporation, R-1061-PR, 1973. As of August 27, 2014:
http://www.rand.org/pubs/reports/R1061.html

Tversky, Amos, and Daniel Kahneman, "Judgment Under Uncertainty: Heuristics and Biases," *Science*, Vol. 185, No. 4157, September 27, 1974, pp. 1124–1131.

U.S. Army Cost and Economic Analysis Center, *Department of the Army Cost Analysis Manual*, Washington, D.C.: Department of the Army, May 2002. As of August 28, 2014:
http://oai.dtic.mil/oai/
oai?&verb=getRecord&metadataPrefix=html&identifier=ADA437381

Varian, Hal R., *Microeconomic Analysis*, 3rd ed., New York: Norton, 1992.

Younossi, Obaid, Mark V. Arena, Robert S. Leonard, Charles Robert Roll Jr., Arvind Jain, and Jerry M. Sollinger, *Is Weapon System Cost Growth Increasing? A Quantitative Assessment of Completed and Ongoing Programs*, Santa Monica, Calif.: RAND Corporation, MG-588-AF, 2007. As of August 28, 2014:
http://www.rand.org/pubs/monographs/MG588.html

Younossi, Obaid, Mark A. Lorell, Kevin Brancato, Cynthia R. Cook, Mel Eisman, Bernard Fox, John C. Graser, Yool Kim, Robert S. Leonard, Shari Lawrence Pfleeger, and Jerry M. Sollinger, *Improving the Cost Estimation of Space Systems: Past Lessons and Future Recommendations*, Santa Monica, Calif.: RAND Corporation, MG-690-AF, 2008. As of August 28, 2014:
http://www.rand.org/pubs/monographs/MG690.html